Trusted Digital Circuits

Hassan Salmani

Trusted Digital Circuits

Hardware Trojan Vulnerabilities, Prevention and Detection

Hassan Salmani
EECS Department
Howard University
Washington, DC, USA

ISBN 978-3-030-07722-8 ISBN 978-3-319-79081-7 (eBook)
https://doi.org/10.1007/978-3-319-79081-7

Softcover re-print of the Hardcover 1st edition 2018

Printed on acid-free paper

This Springer imprint is published by the registered company Springer International Publishing AG part of Springer Nature.
The registered company address is: Gewerbestrasse 11, 6330 Cham, Switzerland

To Maryam and Karim

Preface

The complexity of modern designs, the significant cost of research and development, and the shrinking time-to-market window heavily enforce the horizontal integrated circuit design flow. Many entities across the globe might be involved in the flow and none are necessarily trusted. A malicious party can launch a hardware Trojan attack through manipulating a circuit to undermine its characteristics under rare circumstances at different stages of the flow before and after circuit manufacturing. Detection of hardware Trojans using existing pre-silicon and post-silicon verification techniques is a very challenging task because of the complexity of modern designs, their variety of application, and limited time for verification.

This book is entirely dedicated to study hardware Trojans across the integrated circuit design flow. Unprecedentedly, the book carefully studies integrated circuits at register-transfer level, gate level, and layout level against hardware Trojans. Vulnerabilities of each level to hardware Trojan insertion are discussed, and existing solutions for preventing and detecting hardware Trojans are studied. The book extends its study to hardware Trojan detection after integrated circuit manufacturing and deliberates current testing techniques for hardware Trojan detection. Vulnerabilities of mixed-signal circuits do not remain hidden, and the book studies possible hardware Trojan design in mixed-signal circuits and evaluates existing techniques for hardware Trojan prevention and detection in mixed-signal circuits.

This book is organized into nine chapters. Chapter 1 provides insights into the global integrated supply chain accompanied by statistics. It further defines hardware Trojans and explores their details. Chapter 2 studies vulnerabilities of an integrated circuit at the register-transfer level to hardware Trojan insertion. It comprehensively studies some of promising techniques for vulnerability quantification. In two major parts, Chap. 3 continues with hardware Trojan prevention and detection at the register-transfer level and discusses their effectiveness. Chapter 4 targets the vulnerabilities of gate-level circuits to hardware Trojans. Its main focuses are on techniques suggesting vulnerability quantifications at this level. Chapter 5 presents some of the best known techniques for hardware Trojan prevention and detection in a gate-level circuit. Chapter 6 continues with vulnerabilities of an integrated circuit to hardware Trojan insertion at the layout level. Chapter 7 then discusses some of the known techniques for hardware Trojan prevention and detection at the layout level in detail. Chapter 8 presents an overview on some existing trusted test pattern generation for hardware Trojan detection after integrated circuit manufacturing. Chapter 9, the last chapter, demonstrates hardware Trojan implementation in mixed-signal circuits. It then studies some of the existing techniques for their prevention and detection.

The book offers a comprehensive and detailed analysis of hardware Trojans before and after integrated circuit manufacturing. This book provides design practitioners with guidance on protecting their designs against hardware Trojans and reveals research shortcomings that require attention to address hardware Trojans. The author would like to acknowledge that a part of this book is based on his research during a PhD program under the supervision of Dr. Tehranipoor at the University of Connecticut.

Washington, DC, USA Hassan Salmani

Contents

Chapter 1
The Global Integrated Circuit Supply Chain Flow and the Hardware Trojan Attack

1.1 The Global Integrated Circuit Supply Chain

Integrated circuits (ICs) are the brain of any electronic computing machine that are woven into the fabric of modern living where they deliver accuracy, throughput, and dependability. Their usages have been expanded over the decades from main frames and personal desktop computers to smart wristwatches, autonomous vehicles, and smart homes thanks to Internet of Things (IoT). IoT offers connectivity of electronic devices and enables information-based decision making to provide services in more efficient manners. Figure 1.1 shows revenues from the semiconductor industry market for integrated circuits worldwide from 2009 to 2018. In 2018, revenue from integrated circuit sales is expected to reach US $364.03 billion predicted by Statista [1]. The same source has predicted that the global IoT market will reach more than US $1.7 trillion by 2019, with the number of connected devices worldwide forecast to reach 20.35 billion in the same year.

In spite of the huge market of integrated circuits that has created an intense race between semiconductor vendors, such as Intel and Samsung, the cost of a new design start at 45 nm or below is now approaching (or will even exceed) $50M. Given this research and development (R&D) investment in a single-chip design start, the semiconductor supplier undertaking this design start would need to sustain a significant percentage of a $500M+ market for the return on investment (ROI) on that chip to make sense. In addition to cost, time-to-market plays a critical role in this race. Market windows for new products are becoming short. Therefore, it becomes critical for semiconductor suppliers to ensure that designs arrive the market as early as possible within the target market window in order to maximize the revenues from that design. Otherwise, a significant loss of potential revenue or even a total loss of the market opportunity is plausible. These realities have made the need for design techniques based on intellectual property (IP) reuses (the horizontal integrated circuit design flow) a necessity to amortize these costs over several applications (or generations thereof) in order to get the desired ROI on their initial investment [2]. Further, the companies have outsourced their manufacturing to merchant foundries,

H. Salmani, *Trusted Digital Circuits*, https://doi.org/10.1007/978-3-319-79081-7_1

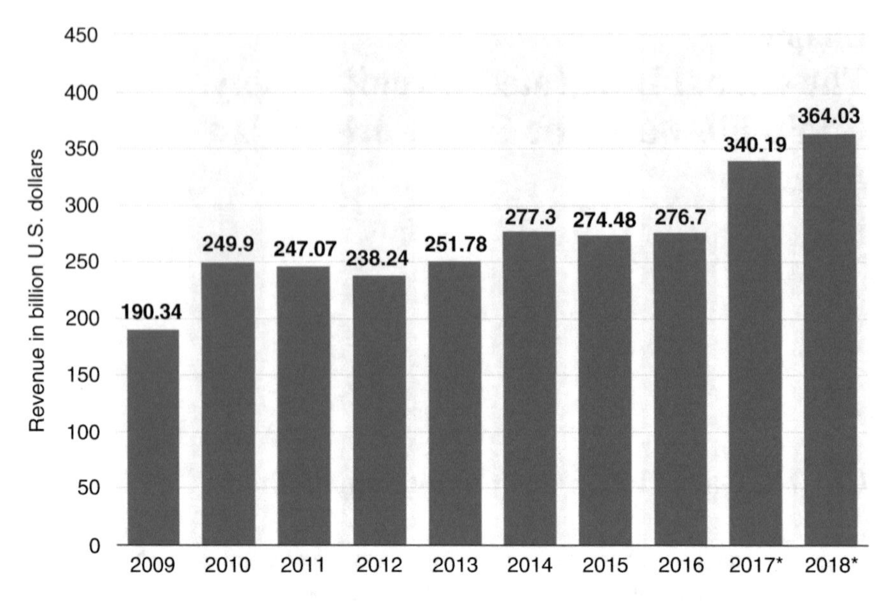

Fig. 1.1 The statistic shows the global revenue from integrated circuits from 2009 to 2018 [1]

and they even further have outsourced the design and verification of their chips to the third-party design service companies. Instead they have focused on core competence like research and development of new technologies and defining protocols [3].

This trend has significantly driven the semiconductor silicon IP market at the rapid pace in the recent years. The market is estimated to change from the value of US $3.306 billion in 2017 to a value of US $6.45 billion by 2022, at a CAGR of 11.71% over the forecast period 2017–2022 [4]. The various applications of the IP market include smart devices, automotive, and computers & peripherals. The major drivers of the market are the emerging consumer devices adoption, demand for connected devices, coupled with the demand for modern system-on-chip (SoC) designs. A SoC is a complete system that is realized on one chip. The "system" mainly consists of a microprocessor, memory, and peripherals. The processor may be a custom or standard microprocessor. There may be multiple processors that are connected through the network-on-chip interconnection. Figure 1.2 presents a typical SoC with Advanced RISC Machine (ARM) processor, dedicated digital signal processing (DSP) processors, and various kinds for input/output ports such as Ethernet and serial lines that are interconnected through the APB and ASB/AHB interconnections.

There are three main categories of IPs [5]: soft, firm, and hard. Soft IP blocks are specified at the register transfer level (RTL) or higher-level descriptions. As a hardware description language (HDL) is process-independent, they are more suitable for digital cores. They are highly flexible, portable, and reusable, but not necessarily optimized in terms of timing and power. Presented at the layout level, hard IP blocks are highly optimized for a given application in a specific process.

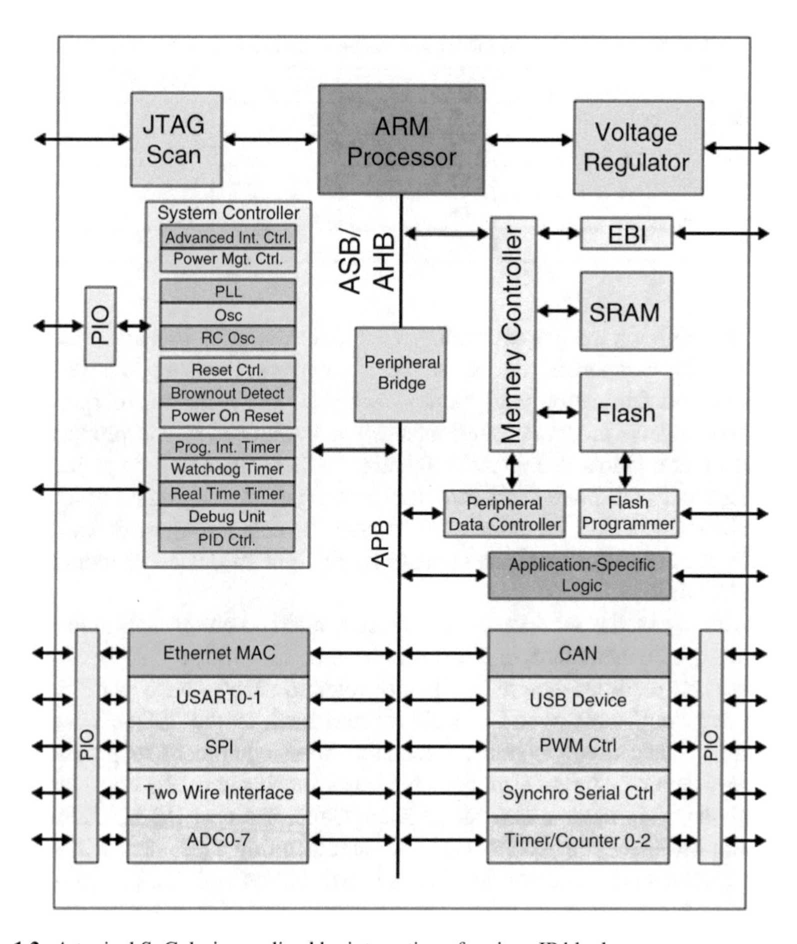

Fig. 1.2 A typical SoC design realized by integration of various IP blocks

Their characteristics are already determined; however, this comes with high cost and lack of flexibility. Firm IP blocks are parameterized circuit descriptions, so they can be optimized according to specific design needs. Firm IPs are between soft and hard IPs, being more flexible and portable than hard IPs, yet more predictable than soft IPs.

Outsourcing causes the emergence of design service companies where there are extensive resources of human intellectuals at a reduced labor cost. For example, the total revenue from very large integrated circuit (VLSI) design services market in India was estimated at US $0.76 billion in 2007 and expected to grow to US $1.38 billion by 2010. Almost 70% of the business comes from the USA while Europe is the next largest contributor. The Indian VLSI design services market comprises of original equipment manufacturers (OEMs), electronic manufacturing

Table 1.1 VLSI projects breakup by the type of design in India between 2007 and 2010 [6]

	2007 (%)	2008 (%)	2009 (%)	2010 (%)
Specification to tape-out	10.9	11.0	12.5	13.8
Module design and verification	41.8	40.4	39.5	38.4
Physical design	24.7	24.3	22.8	21.4
IP development	15.8	16.0	16.1	17.0
Chip testing	6.8	8.3	9.1	9.4

service (EMS), chip design companies, electronic design automation (EDA) tool companies, IP companies, design services companies, testing and verification companies, and fabrication equipment companies. The instances of specification to tape-out projects are in constant increase in India. End-to-end product design will gain more traction as the market matures further. There will be an increase in IP development with more design services moving closer to product development. With an increase in skill set in the testing space, the chip testing market will go up. Table 1.1 presents VLSI projects breakup by the type of design in India between 2007 and 2010 [6].

In early day of the semiconductor industry, a single company would often be able to design, manufacture, and test a new chip. However, the costs of building manufacturing facilities—more commonly referred "fab"—have gone extremely high. A fab could cost over US $200 million back in the 1980s; however, by employing advanced semiconductor manufacturing equipment to produce chips with ever-smaller features, a modern fab costs much more [7]. For example, in late 2012 Samsung made a new fab in Xian, China that cost US $7 billion. It has been estimated that "[i]ncreasing costs of manufacturing equipment will drive the average cost of semiconductor fabs between $15 billion and $20 billion by 2020" [8]. While the congressional interest is the retention of high-value semiconductor manufacturing in the United States, US companies are building semiconductor fabrication plans (fabs) abroad, primarily in Asia. Furthermore, some semiconductor firms are becoming "fab-less," focusing corporate resources on chip design and relying on contract fabs abroad to manufacture their products. At year-end 2015, there were 94 advanced fabs in operation worldwide, of which 17 were in the United States, 71 in Asia (including 9 in China), and 6 in Europe. The Chinese government regards the development of a domestic, globally competitive semiconductor industry as a strategic priority with a stated goal of becoming self-sufficient in all areas of the semiconductor supply chain by 2030. China faces significant barriers to entry in this mature, capital-intensive R&D-intensive industry [9]. It would be also interesting to look at how the foundry market by feature dimension is being predicted till 2025 in Fig. 1.3 [10]. A considerable growth is being predicted for technology nodes less than 10/7 nm. The predication can be interpreted as a higher force behind design outsourcing due to the significant cost of fabs at such low technology nodes.

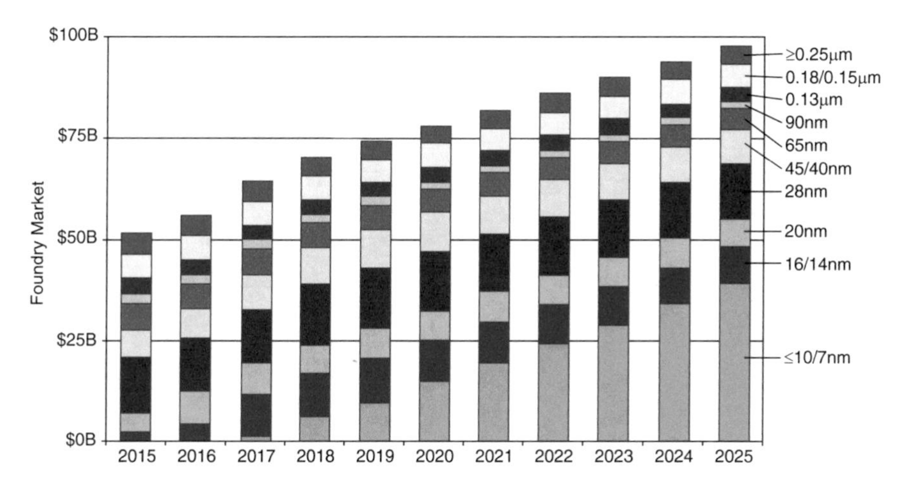

Fig. 1.3 Foundry market by feature dimension [10]

Table 1.2 Dependability attributes

Attribute	Definition
Availability	Readiness for correct service
Reliability	Continuity of correct service
Safety	Absence of catastrophic consequences on the users and the environment
Integrity	Absence of improper system alteration
Maintainability	Ability to undergo modification and repairs
Confidentiality	The absence of unauthorized disclosure of information

1.2 The Hardware Trojan Attack

The dependability of a computer system determines its accountability. The dependability of a system is based on the compliance of delivered services by the system with its functional specifications. The function of the system is described by functional specifications in terms of functionality and performance. The service delivered by the system, on the other hand, is its behavior as it is perceived by its user(s). A broad concept, dependability encompasses availability, reliability, safety, integrity, and maintainability attributes as described in Table 1.2 [11].

Security is more specific, focusing on availability, integrity, and confidentiality. System security demands availability for only authorized actions, integrity with improper meaning unauthorized, and confidentiality. Trust is the dependency of a system (system A) to another system (system B), through which the dependability of system A is affected by the dependability of system B. Trustworthiness in a system is the assurance that the system will perform as expected [11].

A modern society utterly depends on integrated circuits, or chips, which are the virtual brains for all electronics. While economical matters push the global supply chain for integrated circuits, they are becoming increasingly vulnerable

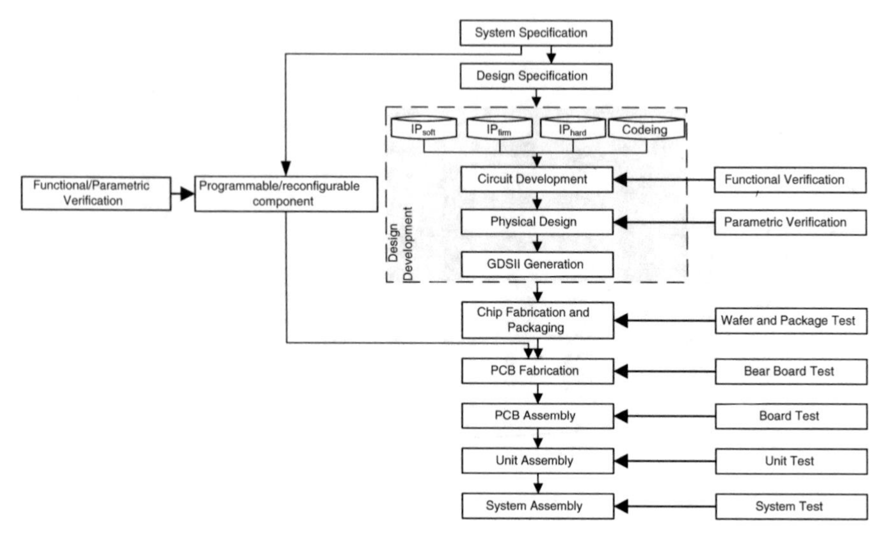

Fig. 1.4 System integration and test process

to malicious activities such as reverse engineering, overproduction, and hardware Trojan insertion. A computer system development, as shown in Fig. 1.4, consists of several steps which are not necessarily performed all in the same design house. The first step is to determine system specifications based on the customer's needs. A complex system may require a variety of components like memories and chips with different applications and functionality.

After providing the system specifications and choosing the structure of system and its required components, design development requires different tools. Each component demands specific attention to meet all the system specifications. To expedite system development and to reduce the final cost, outsourced alternatives have gradually replaced in-house processes. Third-party IP cores have displaced the in-house libraries of logic cells for synthesis. Commercial software has supplanted homegrown computer-aided design (CAD) tool software. In the next step, designed chips are signed off for fabrication. Nowadays, most companies are fabless, outsourcing mask production and fabrication. Beside custom designs, companies can reduce total cost and accelerate system development by using commercial-off-the-shelf (COTS), reprogrammable modules, like microcontrollers, reconfigurable components, or field programmable gate arrays (FPGAs). Afterward, they manufacture printed circuit boards (PCBs) and assemble system components on them. Finally, the PCBs are put together to develop units; the entire system is the integration of these units.

In each step, different verifications or tests are performed to ensure its correctness, as shown in Fig. 1.4. Functional and parametric verifications ascertain the correctness of design implementation in terms of service and associated requirements, like power and performance. Wafer and package tests after the fabrication

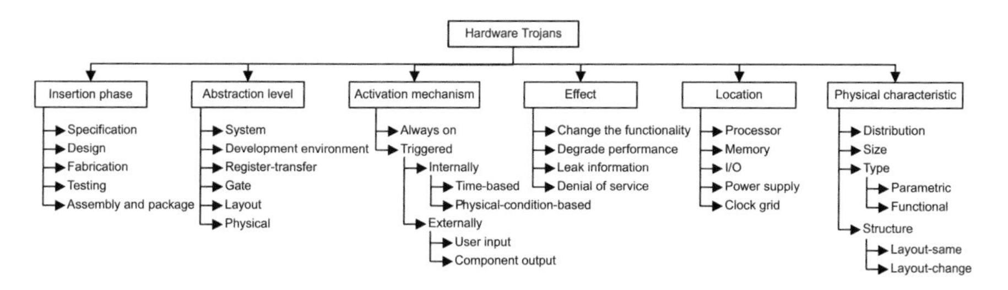

Fig. 1.5 Hardware Trojan Taxonomy [13]

of custom designs separate defective parts and guarantee delivered chips. The PCB fabrication is a photolithographic process and susceptible to defects; therefore, a PCB should be tested before placing devices on it. After the PCB assembly, the PCB is again tested to verify that the components are properly mounted and have not been damaged during the PCB assembly process. The tested PCBs create units and finally the system, which is also tested before shipping for field operation [12].

Each step of system development is susceptible to security breaches. An adversary may change system specifications to make a system vulnerable to malicious activities or susceptible to functional failures. As external resources, like third-party IPs and COTSs, are widely used in design process and system integration, adversaries may hide extra circuit(s) in them to undermine the system at a specific time or to gain control over it. The untrusted foundry issue is rooted in the outsourcing of design fabrication. The untrusted foundry, however, may change the designs by adding extra circuits, like back doors to receive confidential information from the chip, or altering circuit parameters, like wire thickness to cause a reliability problem in the field. The PCB assembly is even susceptible, as it is possible to mount extra components on interfaces between genuine components. In short, a cooperative system development process creates opportunities for malicious parties to take control of the system and to run vicious activities. Therefore, as a part of the system development process, security features should be installed to facilitate trustworthiness, validation, and to unveil any deviation from genuine specifications.

In general, a hardware Trojan is defined as any intentional alteration to a design in order to alter its characteristics. A hardware Trojan has a stealthy nature and can alter design functionality under rare conditions. It can serve as a time bomb and disable a system at a specific time, or it can leak secret information through side channel signals. A hardware Trojan may affect circuit AC parameters such as delay and power; it also can cause malfunction under rare conditions. Figure 1.5 presents the hardware Trojan taxonomy where hardware Trojans are classified based on insertion phase, abstraction Level, activation Mechanism, effect, location, and physical characteristic [13].

In *insertion phase*, a hardware Trojan can be introduced by changing the deign specification, like temperature to degrade the design dependability. Fabrication stages are also subjected to tampering. A hardware Trojan can be realized by adding some extra gates to a design's netlist or by changing its masks. Hardware

Trojan insertion at the testing phase refers to trustworthy testing of a design after fabrication where an adversary may manipulate testing to keep an inserted hardware Trojan undetected. Finally, unprotected interconnections between chips are prone to hardware Trojan interference even if the chips are trustworthy by themselves. An unshielded wire connection could introduce unintended electromagnetic which an adversary can exploit for information leakage or fault injection.

Based on *abstraction level*, the level of abstraction determines the control of advisory on hardware Trojan implementation. At the system level, a design is defined in terms of modules and interconnections between them, with an adversary being limited to the modules' interfaces and their interactions. At the development environment level, a hardware Trojan can be inserted into the modules by taking advantage of CAD tools and scripting languages. In more detail, each module is described in terms of signals, registers, and Boolean functions at the register-transfer level. In this level, an adversary has full access to the functionality and implementation of the modules and can easily change them. A design is represented as a list of gates and their interconnections at the gate level. Here, an adversary can implement hardware Trojans in detail, and hardware Trojans' gates and their interconnections can be decided. At the layout level, the impact of hardware Trojans on design power consumption or its delay characteristics can be controlled. Hardware Trojans can be realized even by changing the parameters of an original circuit's transistors. Finally, all circuit components and their dimensions and locations are determined at physical level. A hardware Trojan can be inserted in white/dead spaces of the design layout with the least impact on the design characteristics.

Based on *activation mechanism*, hardware Trojans may always function, or they become conditionally activated. Always-on hardware Trojans start as soon as their hosting designs are powered on while conditional hardware Trojans seek specific triggers either internally or externally to launch. The internal triggers can be timing-based (a hardware Trojan is activated after certain time), or physical-condition-based (a hardware Trojan is activated by certain events, e.g., specific temperature). The externally triggered hardware Trojans track user inputs or components' outputs, and the hardware Trojans get activated if activation condition(s) are met.

Hardware Trojans can be characterized based on their *effects*. They may change a design's functionality, for example, by modifying the data path of a processor. Hardware Trojans can reduce the design performance or degrade its reliability by changing the design parameters. A hardware Trojan may leak the secret key of a cryptographic processor or can cause the denial of a service for an authorized requested service at specific time.

Based on *location*, any part of a design is potentially subjected to hardware Trojan insertion. A hardware Trojan can be distributed over several parts or tightened in one part. A hardware Trojan can tamper with a processor to obtain control over its controller or data path units. A hardware Trojan in a memory can change stored values or block read/write accesses to the memory. On a PCB including several chips, an inserted hardware Trojan on chips' interfaces can disturb communication. A hardware Trojan can even affect the design power supply and alter current and voltage characteristics. Design delay characteristics can change with interrupting the clock grid by a hardware Trojan. The hardware Trojan can freeze part of clock tree and disable some functional modules.

Based on *physical characteristic*, hardware Trojan physical characteristics represent various hardware manifestations. A hardware Trojan can be a functional or parametric type where functional hardware Trojans are realized by transistors/gates addition or deletion and parametric hardware Trojans by wire thickness or any other design parameter modification. The number of transistors/gates added or removed determines the hardware Trojan size. Hardware Trojan distribution indicates how loose or tight hardware Trojan cells are placed in the physical layout. Hardware Trojan structure refers to possible modification of original physical design for hardware Trojan cells placements.

Hardware Trojan detection is very challenging in the era of automated vehicles, smart cities, and IoTs where sophisticated integrated circuits are used in their hearts to deliver performance, bring accuracy, and provide security. Design verification to ensure its correctness is done before and after its manufacturing. It has been observed that verification of modern designs constitutes a considerable portion of system design, incurring a cost of about 57% of the total project time, on average. While traditional verification has more focused on functional and parametric design specifications, security has become a critical requirement for electronic devices for modern designs that are equipped with the connectivity feature. Unfortunately, security requirements are not well specified even in the first place, and they need to be met for devices that were not originally intended to be connected. As a result, a hardware Trojan can impact devices outside of its embedding device through communication networks [14].

Design verification before manufacturing mainly consists of architecture verification, pre-silicon verification, and emulation FPGA prototyping [14]. The architecture verification focuses on functional parameters of the design, communication protocols among IPs, power and performance management schemes, and security protocols among many others. In this stage, the design goes under verification based on typical use cases of the device to identify parameter values that satisfy the device target in terms of power, performance, area, and security. Pre-silicon verification is performed on individual IPs and on integrated IPs in an SoC with different functionality and requirements. Individual IPs are verified to ensure they function as expected. Most SoC integration verification includes system-level simulation under typical use cases. However, such a simulation is very slow at the register-transfer level (soft IPs) and the fact encourages emulation and FPGA prototyping. Verification by emulation and FPGA prototyping is a bridge between pre-silicon and post-silicon verification. An RTL design is mapped into FPGA and can be executed about hundreds to thousands times faster than an RTL simulator. It should be noted that the controllability and observability of design become limited using emulation and FPGA prototyping and any reconfiguration and remapping may take several hours.

Contrary to decades of study and practice of verification, modern designs have challenged aforementioned verification techniques. One motivation for the global

IC supply chain is the reduction of time to market. This need enforces shrinking verification time while incomplete security analyses leave room for various types of hardware Trojans such as time bombs. Another major issue that is becoming more and more intense is the scalability limitation of existing verification techniques confronting sophisticated SoC designs with tens of various IP cores. Such a limitation eases the implementation of hardware Trojan extended across several IP cores. Another major issue for verification is ambiguities in design specifications including security requirements that are being exacerbated in the era of complex designs and applications. While sometimes the ambiguities are intentional for the sake of intellectual property protection, it leaves considerable opportunities for untrusted downstream entities for hardware Trojan insertion. The ambiguities in design specifications even make it challenging to define a golden reference on which many existing hardware Trojan techniques rely. Modern designs intended for IoT applications contain various analog modules such as sensors and actuators. Verification of analog modules is complex and expensive. And, some researchers have demonstrated hardware Trojans in analog circuits where their manipulations incur none to low design overhead.

Post-silicon validation enables to test a design at its target clock speed and perform nonfunctional testing such as power consumption, physical stress, temperature tolerance, and electrical noise margin. Meanwhile, the controllability and observability of internal activities of a design become significantly limited and any change in the design requires silicon respin (i.e., redesign, validate again, and expensive re-fabrication). Furthermore, the amount of time dedicated to post-silicon verification is limited and governed by the time to the market. Physical reality in functional testing, for example, effects of temperature and electrical noise, negatively impacts post-silicon validation. For example, small extra power consumption becomes masked by process and environmental variations [15]. An untrusted foundry can take advantage of these practical limitations and design hardware Trojans with minimal impacts on design functional and nonfunctional specifications. For example, a hardware Trojan can exist in an IP A, and its goal is to modify some internal signals of an IP B when both IPs are powered on. Detection of such a hardware Trojan could be very challenging during post-silicon verification as the two IPs may not be powered on at the same time due to power management protocols.

1.3 Conclusions

The complexity of modern designs, the diversity of their applications, and the shrinking time to market are all driving the horizontal integrated circuit design flow. To be competitive, different IP vendors are reached for various IPs. Then, provided IPs are integrated by another vendor whose design is finally fabricated by an offshore foundry. The final testing might be even performed by another vendor. This complex flow that is stretched across the globe is executed by all untrusted entities. Unfortunately, existing pre-silicon and post-silicon validation techniques cannot cope with the verification needs of modern designs. The fact does leave extensive opportunities for malicious entities at various levels across the design flow to manipulate a design and undermine its original functional and nonfunctional specifications.

References

1. Statista: forecast of worldwide semiconductor sales of integrated circuits. https://www.statista.com/statistics/519456/forecast-of-worldwide-semiconductor-sales-of-integrated-circuits/. Accessed 22 Jan 2018
2. S. Patel, Changing SoC design methodologies to automate IP integration and reuse. https://www.design-reuse.com/articles/21207/ip-integration-reuse-automation.html. Accessed 22 Jan 2018
3. K.M. Mohan, Outsourcing Trends in Semiconductor Industry, Massachusetts Institute of Technology, 2010
4. Semiconductor silicon IP market. https://www.mordorintelligence.com/industry-reports/global-semiconductor-silicon-intellectual-property-market-industry. Accessed 22 Jan 2018
5. R. Saleh, S. Wilton, S. Mirabbasi, A. Hu, M. Greenstreet, G. Lemieux, P.P. Pande, C. Grecu, A. Ivanov, System-on-chip: reuse and integration. Proc. IEEE **94**(6), 1050–1069 (2006)
6. India Semiconductor and Embedded Design Service Industry (2007–2010), *Market, Technology and Ecosystem Analysis* (India Semiconductor Association, Bangalore, 2008)
7. J. Villasenor, *Compromised By Design? Securing the Defense Electronics Supply Chain* (The Center for Technology Innovation at Brookings, Washington, 2013)
8. B. Johnson, D. Freeman, D. Christensen, S.T. Wang, Market trends: rising costs of production limit availability of leading-edge fabs. https://www.gartner.com/doc/2163515/market-trends-rising-costs-production. Accessed 22 Jan 2018
9. M.D. Platzer, J.F. Sargent Jr., U.S. semiconductor manufacturing: industry trends, global competition, federal policy (2016)
10. H. Jones, Whitepaper: semiconductor industry from 2015 to 2025. http://www.semi.org/en/node/57416. Accessed 22 Jan 2018
11. A. Avizienis, J.-C. Laprie, B. Randell, C. Landwehr, Basic concepts and taxonomy of dependable and secure computing. IEEE Trans. Dependable Secure Comput. **1**(1), 11–33 (2004)
12. L.-T. Wang, C.-W. Wu, X. Wen, *VLSI Test Principles and Architectures* (Morgan Kaufmann Publishers, San Francisco, 2006)
13. H. Salmani, M. Tehranipoor, R. Karri, On design vulnerability analysis and trust benchmarks development, in *2013 IEEE 31st International Conference on Computer Design, ICCD 2013*, Asheville, NC, 6–9 October 2013, pp. 471–474
14. W. Chen, S. Ray, J. Bhadra, M. Abadir, L.C. Wang, Challenges and trends in modern SoC design verification. IEEE Des. Test **34**(5), 7–22 (2017)
15. P. Mishra, R. Morad, A. Ziv, S. Ray, Post-silicon validation in the SoC era: a tutorial introduction. IEEE Des. Test **34**(3), 68–92 (2017)

Chapter 2
Circuit Vulnerabilities to Hardware Trojans at the Register-Transfer Level

2.1 Circuits at the Register-Transfer Level

At the register-transfer level, a hardware description language (HDL), such as VHDL or Verilog, is used to describe a circuit by concurrent algorithms (behavioral). There are generally three different approaches to writing HDL architectures: the dataflow style, the behavioral style, and the structural style.

In date flow style, a circuit is specified as a concurrent representation of the flow of data through the circuit. This style is preferred for small and primitive circuits. The behavioral style models how the circuit outputs will react (or behave) to the circuit inputs. In comparison to the dataflow style, the behavioral style provides no details as to how the design is implemented in actual hardware. The structural style is used to manage and organize the implementation of a complex system. In the structural style, a system is expressed in terms of subsystems which are interconnected by signals, for example, a system consisting of RAM, CPU, and ROM. Each subsystem may in turn be composed of a collection of subsubsystems, for example, a CPU consisting of ALU and Register bank, and so on. Until it finally reaches a level only consisting of primitive components, described purely in terms of their behavior. The top-level system can be thought of as having a hierarchical structure.

Circuits at the register-transfer level may not necessarily show how their final implementation at the gate level after the synthesis would look like as the register-transfer-level circuits are technology independent. There is a high flexibility in implementing a circuit at the register-transfer level. For given design specifications, two designers may provide two different implementations using supporting constructs by a specific language while their input–output relationships match. Such flexibility makes the security evaluation of a provided register-transfer-level circuit against hardware Trojans very challenging.

H. Salmani, *Trusted Digital Circuits*, https://doi.org/10.1007/978-3-319-79081-7_2

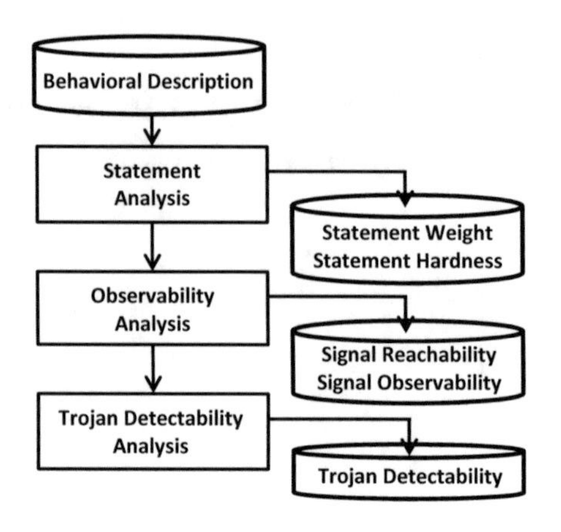

Fig. 2.1 Vulnerability analysis flow at the register-transfer level

2.2 Value Range Analyses for Vulnerability Assessments

Figure 2.1 shows a proposed circuit vulnerability analysis flow to hardware Trojan insertion at the register-transfer level in [1]. The data and control flows of the circuit determine the difficulty of executing a statement and the observability of internal signals. *Statement Analysis* in Fig. 2.1 measures statement weight and statement hardness of a statement which quantify conditions under which the statement executes. In the following, *Observerability Analysis* evaluates reachability of signals from each other and determines their observability through circuit primary outputs. A circuit at the register-transfer level is vulnerable to hardware Trojan insertion where statement hardness is high or observability is low. A hardware Trojan at the register-transfer level (also called "a behavioral hardware Trojan") can change a statement rarely executed or carry out an attack through a signal with very low observability. *Hardware Trojan Detectability Analysis* gauges the detection of a malicious change of a statement or the inclusion of a new statement based on its weight and observability of its target signal.

2.2.1 *Statement Analysis*

The proposed vulnerability analysis at the register-transfer level quantifies (1) statement hardness for each circuit statement and (2) observability for each circuit signal, based on a weighted value range. Adopted from the work presented in [2] to improve the accuracy of static value and branch prediction in compilers, the notation $\{W[L, U]\}$ is developed for each signal where W represents the weight of the value range, and L and U show the lower and upper limits of the value range. Circuit control and dataflows determine W, L, and U for each statement. The technique creates a condition stack to track the circuit control flow. Further, it generates an individual stack for each signal to pursue the circuit dataflow. While the circuit code

```
1. LIBRARY IEEE;
2. USE IEEE.STD_LOGIC_1164.ALL;
3. ENTITY EXAMPLE IS
4. PORT( CLK : IN BIT;
5.          X : IN INTEGER RANGE 15 DOWNTO 0;
6.          K : IN INTEGER RANGE 15 DOWNTO 0;
7.          Z : OUT INTEGER RANGE 15 DOWNTO 0);
8. END ENTITY EXAMPLE;
9. ARCHITECTURE SIMPLE OF EXAMPLE IS
10. BEGIN
11. PROC1: PROCESS (CLK)
12. VARIABLE P : INTEGER RANGE 15 DOWNTO 0;
13. BEGIN
14.         FOR X IN 0 TO 9 LOOP
15.                 IF ( X < 2 ) THEN
16.                         P := 1 - X;
17.                 ELSIF ( X > 5 ) THEN
18.                         IF ( K = 7) THEN
19.                                 P := 2 * X;
20.                         ELSE
21.                                 P := 2 + K;
22.                         END IF;
23.                 END IF;
24.         END LOOP;
25.         IF ( P < 7 ) THEN
26.                 IF ( X < 7 ) THEN
27.                         IF ( P > 2 ) THEN
28.                                 Z <= P;
29.                         END IF;
30.                 END IF;
31.         END IF;
32. END PROCESS PROC1;
33. END ARCHITECTURE SIMPLE;
```

Fig. 2.2 A sample register-transfer-level code

is being parsed, observing a condition block limiting the value range of a signal, such as loop and condition statements, pushes a new condition into the condition stack and a new weighted value range into the stack of the signal. Reciprocally, exiting a condition block pops from the condition stack and may update the stacks of signals.

A weighted value range is determined by a condition or assignment statement. The weight of a range is defined as $\left(\frac{U-L+1}{U_O-L_O+1}\right)$, where U and L are the upper and lower limits of the controlling signal at the top of the condition stack, and U_O and L_O are the declared upper and lower limits of the controlling signal. If different assignment statements to one target signal take place under different conditions in one condition block, the weighted value ranges of the target signal will merge when exiting the condition block. Further, the statement hardness is defined as the reverse of the weight of the controlling signal of that statement (*the statement weight*).

The technique was applied to a small register-transfer-level code, shown in Fig. 2.2, and the statement hardness and the weighted value range of controlling signals X, K, and P are presented in Table 2.1. Further, Fig. 2.3 shows the condition stack and stacks for the controlling signals over the code execution.

The program in Fig. 2.2 consists of one sequential block between Lines 13 and 32. In Line 13, where the block begins, the condition stack is empty. The controlling

Table 2.1 Statement hardness and value range of controlling signals for the code in Fig. 2.2

Line number	Statement hardness ($1/T_W$)	Value range		
		X	K	P
13	1	{1[0, 15]}	{1[0, 15]}	{1[0, 15]}
14	1	{1[0, 15]}	{1[0, 15]}	{1[0, 15]}
15	1/0.625	{0.625[0, 9]}	{1[0, 15]}	{1[0, 15]}
16	1/0.125	{0.125[0, 1]}	{1[0, 15]}	{0.125[0, 1]}
17	1/0.625	{0.625[0, 9]}	{1[0, 15]}	{0.125[0, 1]}
18	1/0.25	{0.25[6, 9]}	{1[0, 15]}	{0.125[0, 1]}
19	1/0.015625	{0.25[6, 9]}	{0.25[7, 7]}	{0.015625[12, 15]}
20	1/0.25	{0.25[6, 9]}	{1[0, 15]}	{0.015625[12, 15]}
21	1/0.234375	{0.25[6, 9]}	{0.25[0, 6], 0.25[8, 15]}	{0.234375[2, 8], 0.234375[10, 15]}
22	1/0.25	{0.25[6, 9]}	{1[0, 15]}	{0.25[2, 8], 0.25[10, 15]}
23	1/0.625	{0.625[0, 9]}	{1[0, 15]}	{0.625[0, 8], 0.625[10, 15]}
24	1	{1[0, 15]}	{1[0, 15]}	{0.625[0, 8], 0.625[10, 15]}
25	1	{1[0, 15]}	{1[0, 15]}	{0.625[0, 8], 0.625[10, 15]}
26	1/0.4375	{1[0, 15]}	{1[0, 15]}	{0.4375[0, 6]}
27	1/0.19140625	{0.19140625[0, 6]}	{1[0, 15]}	{0.4375[0, 6]}
28	1/0.0478515625	{0.19140625[0, 6]}	{1[0, 15]}	{0.0478515625[3, 6]}
29	1/0.19140625	{0.19140625[0, 6]}	{1[0, 15]}	{0.4375[0, 6]}
30	1/0.4375	{1[0, 15]}	{1[0, 15]}	{0.4375[0, 6]}
31	1	{1[0, 15]}	{1[0, 15]}	{0.625[0, 8], 0.625[10, 15]}
32	1	{1[0, 15]}	{1[0, 15]}	{0.625[0, 8], 0.625[10, 15]}

			K_{18}	K_{20}						P_{27}			
	X_{15}	X_{17}	X_{17}	X_{17}	X_{17}				X_{26}	X_{26}	X_{26}		
X_{14}	X_{14}	X_{14}	X_{14}	X_{14}	X_{14}	X_{14}		P_{25}	P_{25}	P_{25}	P_{25}	P_{25}	
Line 14	Line 15	Line 17	Line 18	Line 20	Line 22	Line 23	Line 24	Line 25	Line 26	Line 27	Line 29	Line 30	Line 31

Fig. 2.3 The control flow for the code in Fig. 2.2

signals are set to their full range with the weight of 1. The first statement in Line 14 executes unconditionally and its statement hardness is set to 1. The loop in Line 14 limits the controlling signal X between 0 and 9, so $L = 0$ and $U = 9$. While $L_O = 0$ and $U_O = 15$, as declared in Line 5, the weight of the value range is $\left(\frac{9-0+1}{15-0+1} =\right) 0.625$. Therefore, the hardness of the statement in Line 15 is $\frac{1}{0.625}$. Represented as X_{14} in the condition stack, where the subscripted 14 shows the line number, the signal X is pushed to the condition stack in Line 14 and $\{0.625[0, 9]\}$ into the stack of signal X.

The immediate condition statement in the next line, Line 15, again pushes the signal X (X_{15}) into the condition stack and pushes the new weighted value range $\{0.125[0, 1]\}$ into the stack of signal X. The hardness of Line 16 is $\frac{1}{0.125}$ and the assignment statement for this line defines a new weighted value range for the signal P; the weighted value range, $\{0.125[0, 1]\}$, is then pushed into the stack of signal P. The ELSE statement in Line 17 in Fig. 2.2 pops from the condition stack (X_{15}) and from the stack of signal X. Afterward, the top element of the condition stack is X_{14}, so the statement hardness of Line 17 is $\frac{1}{0.625}$. Further, the ELSE statement in Line 17 pushes X (X_{17}) into the condition stack, generates a new weighted value range $\{0.25[6, 9]\}$, and pushes into the stack of signal X.

In the following, the hardness of executing statements in Line 18 is $\frac{1}{0.25}$. In Line 18, a new condition block is defined over K inside the condition block already defined in Line 17. The last condition statement limits the controlling signal K to 7, and the new weighted value range of K, that is, $\{0.25[7, 7]\}$, is pushed into the stack of signal K and K_{18} into the condition stack. By executing the assignment statement in Line 19, a new weighted value range $\{0.015625[12, 15]\}$ is pushed into the stack of signal P. The ELSE condition statement in Line 20 pops K_{18} from the condition stack and removes the top range of the stack of signal K. Afterward, the top controlling signal is X_{17} that determines the statement hardness of Line 20. Then, the new value range of the signal K, $\{0.25[0, 6], 0.25[8, 15]\}$, is pushed into the stack of signal K. The rest of code can be analyzed in the same manner.

A signal's weighted value range indicates the range of values the signal can take. Constrained by circuit data and control flows, the range may only cover part of the original declared range of the signal. The weight of the range, determined by statement hardness, represents the difficulty of assigning the range to the signal. Table 2.2 shows the weighted value range of the output signal Z and the internal signal P for the code in Fig. 2.2.

The declared value range of P and Z is $[0, 15]$. However, the value range of the signal P considering the circuit control and dataflows does not include

Table 2.2 The value range of internal and output signals for the code in Fig. 2.2

Signal name	Value range
P	{0.625[0, 8], 0.625[10, 15]}
Z	{0.0478515625[3, 6]}

the value 9. Similarly, Table 2.2 shows that the output signal Z only covers the value range of [3, 6] with comparatively low weight, and the signal does not cover 75% of its declared range. Uncovered value ranges can be potentially used for behavioral hardware Trojans implementation because circuit verification and the normal operation of the circuit do not reach values out of range. Knowing that, an attacker may design a hardware Trojan to propagate unexpected values and initiate a malfunction.

Independent from circuit input vectors, statement hardness provides a quantitative measure of the difficulty of executing a statement. A statement with a high level of hardness is rarely executed and its correctness cannot be fully examined with applying a limited number of test input vectors in a testbench form. In the example above, the assignment statement in Line 19 with a statement hardness of (1/0.015625=)64 is the most difficult statement in the code to be exercised due to nested conditional statements which control the execution of the statement. Therefore, modifications to the statement or the inclusion of new statements under the same condition will be hard to detect. This analysis reveals parts of a circuit that are more vulnerable to hardware Trojan insertion. As another measure, the observability of a signal characterizes the difficulty of observing the signal through a circuit primary output.

2.2.2 *Observerability Analysis*

A behavioral hardware Trojan usually targets a signal with low observability to carry out an attack. The vulnerability analysis technique develops a weighted data graph of a circuit while it parses circuit code in order to calculate observability along with statement hardness. The graph shows the connectivity of signals that appeared in code. Figure 2.4 shows the data graph of the code in Fig. 2.2.

The nodes of the data graph are circuit signals, and directed edges show their dependency. Edges are weighted with the sum of the weights of assignment statements that represent the source node as an input and the destination node as the target signal. For example, the signal X appears in two assignment statements in Lines 16 and 19 where the target signal is the signal P in both. The weights of the assignment statements, as shown in Table 2.1, are 0.125 and 0.015625, respectively; therefore, the weight of the edge from the signal X to the signal P is $(0.125 + 0.015625 =)0.140625$. In another case, the weight of the edge from the signal K to the signal P is the weight of the assignment statement in Line 21,

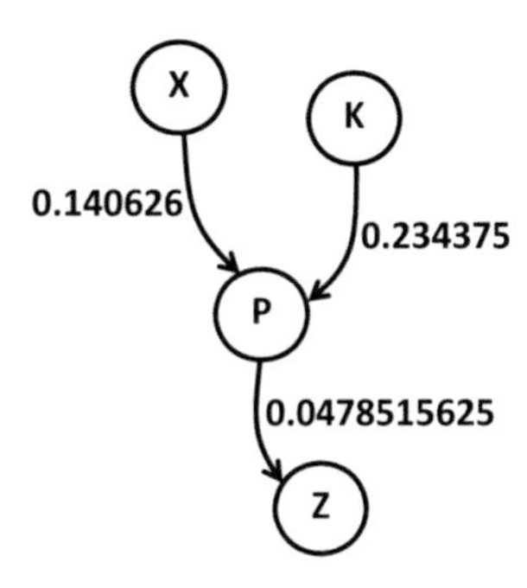

Fig. 2.4 The data graph for the code in Fig. 2.2

Table 2.3 The reachability and observability of signals for the code in Fig. 2.2

Signal name	Target signal	Reachability	Observability(T_O)
X	Z	0.006729	0.006729
X	P	0.140625	0
P	Z	0.047851	0.047851
K	Z	0.011215	0.011215
K	P	0.234375	0

which is 0.234375, as shown in Table 2.1. Based on the data graph, it is possible to calculate the reachability of signals from each other and the observability of signals at circuit outputs. Table 2.3 shows the reachability and observability of signals in the circuit in Fig. 2.2, based on the data graph in Fig. 2.4.

The vulnerability analysis flow at the register-transfer level determines the reachability of signals that can directly or indirectly reach each other. With the assignment statement in Line 28, the signal P directly reaches the signal Z with a reachability of 0.047851, which is the weight of edge from P to Z. As the signal Z is an output signal, the observability of H is also 0.047851.

The signal X reaches the signal P, and its reachability is 0.140625. But its observability is 0 as the signal P is not an output signal. However, the signal X is observable through the signal Z, an output signal. The reachability of the signal X to the signal Z is $(0.140625 \times 0.047851 =)0.006729$, and with the same value, it is observable. In a similar manner, the reachability and observability of signal pairs can be calculated. The signal P has the highest observability, as it is directly assigned to the output signal Z. On the other hand, the signal X is only observable with the minimum value of 0.006729 through the output signal Z.

Reachability of a signal pair signifies driving strength of a source signal in determining the final value of a destination signal. As shown in Fig. 2.4, the signal P is driven by the signals X and K. The reachability of the pair K and P is 0.234375 and that of the pair X and P is 0.140626, that is, the signal K has more contribution in the final value of the signal P. Confirmedly Table 2.1 indicates the signal K determines the widest range of the signal P with a higher weight compared with the signal X which derives the signal P in Lines 16 and 19 with smaller ranges and lower weights.

The vulnerability analysis flow determines the statement hardness of each statement in a code. The statement hardness exposes parts of the code which are more vulnerable to hardware Trojan insertion. Further, it is possible to obtain the weighted value range of circuit signals at any line of the code. This information can be used to determine the root of high statement hardness. The flow makes it also possible to calculate the reachability of signals from each other and their observability by an output signal. The reachability analysis determines to what extent a signal can impact a target signal, and the observability analysis indicates how difficult it is to monitor an internal signal by an output signal.

The vulnerability analysis flow consists of two main steps. In the first step, as a code is being parsed, the hardness of each statement is calculated and a data graph is developed. In the following step, the reachability and observability of signals are calculated using the data graph. The complexity of the first step is $O(L)$, where L is the number of statements. Suppose S is the number of signals; the complexity of the second step would be $O(S \times S)$. Therefore, the complexity of the vulnerability analysis algorithm is $O(L) + O(S^2)$. It is expected that the number of signals in a circuit is much less than the number of statements ($S \ll L$), thus the complexity of the vulnerability analysis algorithm is approximated to $O(L)$.

2.2.3 *Application of Value Range-Based Vulnerability Analysis*

The proposed vulnerability analysis is applied to several benchmarks, and in this section results on b01, b02, b03, b04, and b05 of ITC99's [3], and SAP benchmarks [4] are presented. Table 2.4 demonstrates the details of each benchmark. As there is no unique way of implementing a circuit at a high level of abstraction (i.e., here, register-transfer level), we modified the circuits slightly to prepare for analysis using our automated flow, while keeping the original functionality intact.

The results for vulnerability analysis for the benchmarks are presented in Table 2.5. The minimum statement hardness of the five benchmarks is 1, which means the corresponding statement is executed unconditionally. Nested conditional statements increase statement hardness. The highest statement hardness of b01

Table 2.4 Benchmarks characteristics

Name	# VHDL Lines	PI	PO	Function
b01	115	4	2	FSM
b02	55	3	1	IsBCD
b03	175	6	4	Arbiter
b04	93	13	8	MinMax
b05	278	3	6	Memory contents elaboration
SAP	396	2	2	8-bit 8080-based microprocessor

Table 2.5 Vulnerability analysis for the benchmarks

	Statement hardness		Observability	
Name	Min	Max	Min	Max
b01	1	1/0.015625	0	0.5
b02	1	1/0.035714286	0	0
b03	1	1/0.005208333	0	0.166
b04	1	1/0.041666667	0	0.25
b05	1	1/2.26E−06	0	0
SAP	1	1/0.002403846	0	0.25

is (1/0.015625 =) 64, where the corresponding statement is executed when four conditions are met. b01 has one internal signal which is not observable through any of its outputs; therefore, the minimum observability for b01 is 0. Two primary inputs are observable through one of its outputs and determine the maximum observability of b01, which is 0.5.

b02 benchmark is the smallest circuit, and its minimum and maximum statement hardnesses are 1 and (1/0.035714286 =) 28, respectively. The only output of the circuit is assigned constant values, so no internal signals or primary inputs are observable. Hence, the minimum and maximum observability for b02 are 0.

None of primary inputs of b03 is observable through its primary outputs; therefore, the minimum observability is 0. The only internal signal that is assigned constant values is observable with the maximum observability of 0.1666. The maximum statement hardness of b03 is (1/0.005208333 =) 192, and the corresponding statement is executed when four conditions are met.

All internal signals of b04 circuit are observable only through its primary output, and b04 has the highest maximum observability with value of 0.25. The minimum observability is 0, which belongs to some primary inputs. The maximum statement hardness of b04 is (1/0.041666667=) 24, whose corresponding statement is executed when four conditions are met.

b05 has the highest maximum statement hardness among all with the value of (1/2.26E−06 =) 442,477. The considerable high statement hardness is attributed to the deep nested condition blocks. The corresponding statements are executed when 11 conditions are met. The minimum and maximum observability of b05 are 0 as all outputs are assigned to constant values. SAP is the largest circuit with the maximum statement hardness of (1/0.002403846 =) 416, and with the minimum and maximum observability of 0 and 0.25, respectively.

In above circuits, signals with 0 observability mainly serve as condition signals for concurrent or sequential conditional and selected signal assignments. Therefore, they cannot be directly observed through any primary output.

Figure 2.5 shows the distribution of statement weight for the benchmarks. "Normalized Frequency" in the figure indicates the number of lines with a specific statement weight in a benchmark divided by the number of VHDL lines inside the benchmark's VHDL process statements. As soon, a portion of statements in the circuit presents very low statement weight. Assuming that any statement with

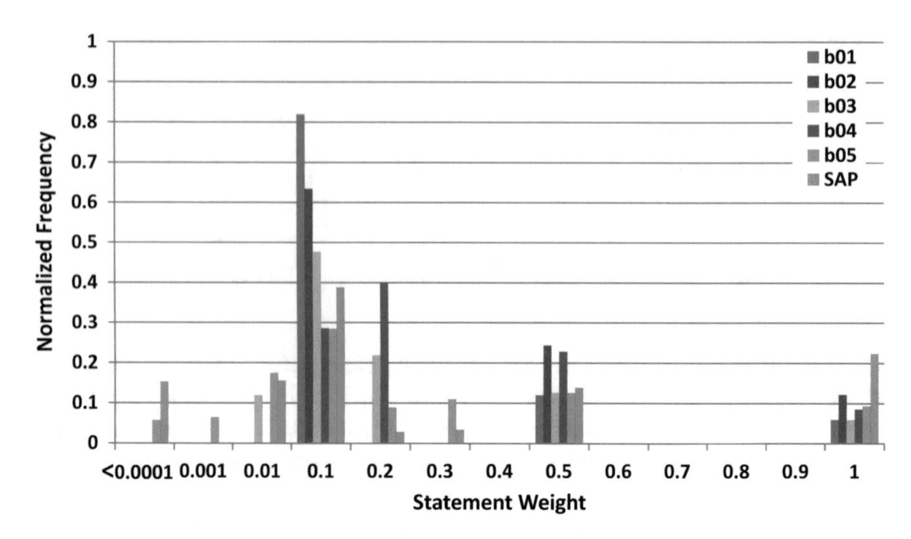

Fig. 2.5 Statement weight analysis

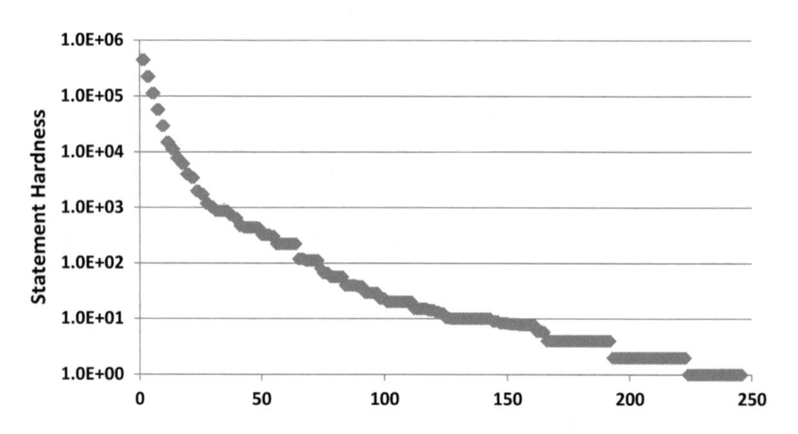

Fig. 2.6 Statement hardness for b05

statement weight lower than 0.0001 is a rarely executed statement, one can further analyze these statements and generate test benches to activate them and analyze the responses.

The statement hardness analysis for b05 shown in Fig. 2.6 presents the number of lines with significant large statement hardness ranging from 1.13E+04 to 4.42E+05. To prevent hardware Trojan insertion in a circuit at the register-transfer level, it is possible to define a specific threshold on statement weight (statement hardness) and modify the circuit such that the statement weight (statement hardness) of all statements stands above (below) the threshold.

The detail analysis reveals the statement hardness depends on both the value range and the control flow. Defining a condition block on signals with a wide value range increases the statement hardness of statements inside the condition block.

Nested condition blocks may increase statement hardness of interior statements as well. In above examples, although the statements with the highest statement hardness in b01 and b03 are both executed when four conditions are met, the execution of the statement in b03 is limited over the larger value range, which increases its statement hardness 3 times. On the other hand, for b04 benchmark, it is observed that the statement with the highest statement hardness is executed when five conditions are met, while its statement hardness is the minimum among all the benchmarks due to the limited value range.

The significance of this vulnerability analysis is that it determines which parts of a circuit are hard to activate without needing a testbench. While developing a testbench to examine all control and data paths is a challenging task by itself, this vulnerability analysis is based on weighted value range. The weighted value range exactly determines the range of values under which a statement executes. This information can be used to develop a testbench intended to exercise potential hardware Trojan locations.

2.3 Unspecified IP Functionality

Verification and testing of modern designs is a major bottleneck in hardware design, and it is estimated that over 70% of hardware development resources are consumed by the verification task [5]. This situation is more exacerbated by the fact that commercial verification tools have primarily focused on increasing confidence in the correctness of *specified functionality* from security perspective. Behavior which is not modeled will not be verified by existing methods, meaning any security vulnerabilities occurring within unspecified functionality will go unnoticed [6]. An attacker can augment circuit functionality by modifying circuit behavior out of design expected specifications while any maliciously added functionality remains from existing verification methods.

As an example, Fig. 2.7 shows a malicious circuitry shown in red whose operation is hidden in the unspecified specification of the first-in-first-out (FIFO) circuit: "how should the circuit behave if the signal read_data is deasserted?" The

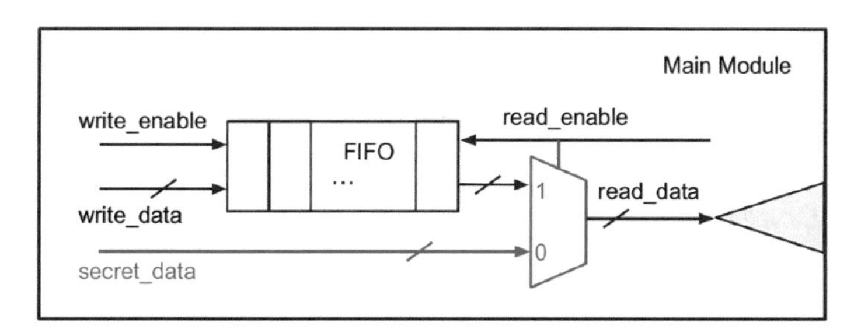

Fig. 2.7 FIFO augmented by adding a hardware Trojan shown in red that operates in the unspecified functionality [7]

FIFO executes two main operations: (1) it stores the input write_data into the FIFO if the signal write_enable is '1' and the FIFO is not full, and (2) puts data from the FIFO on the output read_data when the signal read_enable is '1' and the FIFO is not empty. In this implementation, the input write_data and the output read_data are two different signals, and they can be controlled separately. In a normal store operation, the input read_enable is deasserted while the input write_enable is asserted to store the input write_data. The malicious hardware Trojan in red in Fig. 2.7 can then leak the signal secret_value on the output read_data, as no verification scenario may pay attention to the value of the output read_data when the input write_enable is '1' and the input read_enable is '0'.

2.3.1 Hardware Trojans in Don't Cares

Design specification may only determine how a circuit should operate under certain combinations of inputs. Therefore, it is implicitly assumed any other input vector will never be applied to the circuit. The outputs of a circuit to unexpected input vectors are considered don't cares, and a behavioral synthesizer can take advantage of don't cares to maximally optimize the outcome logic.

In HDL languages, such as Verilog, don't cares can be literally expressed using the literal "X," and a variable assigned "X" can be assigned '1' or '0' by the synthesizer. While don't cares are being extensively used for logic optimization, they indeed introduce undetermined zones where a hardware Trojan can be inserted without violating any expected behavior from a circuit.

Figure 2.8 presents a circuit with one counter and two state machines. One state machine (Lines 21–27) is controlled by the counter and the other state machine (Lines 32–35) is controlled through a primary input. While the counter is a 3-bit counter and can vary between 0 and 7, it just counts between 0 and 3. As a result, the Line 27 is not ever reachable. Line 26 is an assignment to the 4-bit internal signal *pattern*. Bits 3 and 0 of the signal pattern are determined but bits 1 and 2 are defined don't cares. At the first glance, both bits can be used to leak certain data through the only primary output *out*. A detailed analysis shows that only possible leakage can occur through the intermediate signal *tmp* in Line 35 when it is coincident with the counter that becomes 4. As a result, only the bit 2 of pattern is observable from the outside of module. A vulnerability analysis tool should distinguish between Xs and identify those that can subvert the security of design.

2.3.2 Dangerous Don't Cares Identification

A don't care (dc_i) is considered dangerous if the statement assigning dc_i to a design variable is reachable (it is not necessarily the variable to be accessible from a primary input), and the value of the variable must propagate to a circuit output. The

```
1  module simple_state ( clk, reset, control, data, key, out ) ;
2  input clk, reset ;
3  input [1:0] control ;
4  input [3:0] data, key ;
5  output reg [3:0] out ;
6  reg [3:0] tmp ;
7  reg [2:0] counter , next_counter ;
8  reg [3:0] pattern ;
9  // Truncated Counter 0-4
10 // 5, 6, and 7 never appear
11 always @ (*) begin
12     if ( counter < 3'h4 )
13         next_counter <= counter + 3'b1 ;
14     else next_counter <= 3'b0 ;
15 end
16 always @( posedge c lk ) begin
17     if (~ reset ) counter <= 3'b0 ;
18     else counter <= next_ counter ;
19 end
20 always @ (*) begin
21     case ( counter )
22         3'd0:pattern <= 4'b1010 ;
23         3'd1:pattern <= 4'b0101 ;
24         3'd2:pattern <= 4'b0011 ;
25         3'd3:pattern <= 4'b1100 ;
26         3'd4:pattern <= 4'b1xx1 ;
27         default:pattern <= 4 'bxxxx ;
28     endcase
29 end
30 always @ (*) begin
31     case ( control )
32         2'b00:tmp <= data ;
33         2'b01:tmp <= data ^ key ;
34         2'b10:tmp <= ~ data ;
35         2'b11:tmp <= data ^ { pattern [3] , pattern [2:0] & counter } ;
36     endcase
37 end
38 always @ (posedge clk ) begin
39     if (~reset ) out <= 4'b0 ;
40     else out <= tmp ;
41 end
42 endmodule
```

Fig. 2.8 A simple state machine [7]

problem of finding if such an input sequence exists has been formulated in [7] as a sequential equivalence checking problem. Then, to determine a specific reachable dc_i is dangerous, the equivalence check is performed between two near-identical versions of the design: one version with $dc_i = 0$ and the other version with $dc_i = 1$. If the designs are identical under all possible input sequences, dc_i cannot possibly be used to leak design information.

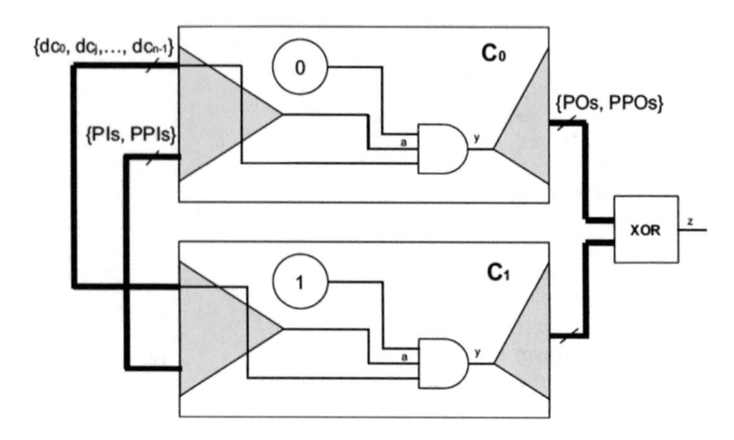

Fig. 2.9 Equivalence checking formulation [7]

To study the relationship between multiple don't cares in the design, the problem is formulated in terms of state reachability analysis and combinational equivalence checking in [8], and the X-analyzer tool is being developed. State reachability analysis is performed first to distinguish ones that are reachable through a sequence of inputs. Afterward, combinational equivalence checking is performed on the set of reachable don't care. To do so, all flop-flops in a sequential circuit are removed and their data inputs of flip-flops are defined as pseudo primary outputs (PPOs) and their data outputs of flip-flops are defined as pseudo primary inputs (PPIs). In the following, combinational equivalence checking is performed between two versions of the original design: $Cdc_i = 0$, and $Cdc_i = 1$, by constructing the miter as shown in Fig. 2.9 and checking the satisfiability of node z. If z is not satisfiable, then dc_i is safe. Otherwise, the equivalence checker returns a distinguishing input vector. Note that when analyzing dc_i, all remaining $n-1$ don't care bits are made primary inputs.

The X-analyzer tool is being applied on Elliptic Curve Processor (ECP) which performs the point multiplication operation optimized for an FPGA implementation [9]. The report of X-analyzer tool is presented in Table 2.6. The ECP design has 572 primary input bits, 467 primary output bits, and 11,232 state elements, resulting in a gate count over 300,000. There are 538 don't care bits in the design analyzed by the tool. 282 correspond to assignments made during states 0–38 to bits in *cwl* and *cwh*, 33 correspond to the *default* assignments (state >38, which should be unreachable) of these signals, and 233 are from a *default* assignment in the *quadblk* module.

2.4 Formal Verification and Coverage Analysis for Vulnerability Analyses

Coverage analyses and functional analyses are used to determine vulnerable portions of a register-transfer level circuit through exhaustively examining all defined functionality in design specifications [10]. All functionality are translated

Table 2.6 Classification of don't cares [7]

Row #	# of don't care bits	Signal(s) affected
Class 1: definitely dangerous (32 bits)		
1	2	$cwh[4]$, $cwh[7]$, when $state == 15$
2	1	$cwh[12]$, when $state == 2$
3	32	cwl, for various state ≤ 38
Class 2: possibly dangerous (272 bits)		
4	6	$nextstate[5:0]$, when $state > 38$
5	23	$cwh[22:0]$, when $state > 38$
6	10	$cwh[9:0]$, when $state > 38$
7	233	$d[232:0]$, when $cwh[19:16] == 1$ or $cwh[19:16] == 15$
Class 3: definitely safe (231 bits)		

to properties (assertions) and coverage metrics are used to identify suspicious parts in a circuit under authentication. Coverage metrics include the code coverage and the functional coverage. While covering various types of analyses, code coverage toward hardware Trojan covers line, statement, toggle, and finite state machine (FSM) coverage. The functional coverage is the determination of how much functionality of the design has been exercised by the verification environment.

If all the assertions generated from the specification of a circuit are successful and all the coverage metrics such as line, statement, and FSM are 100%, then it can be assumed with high confidence that the circuit is hardware Trojan-free. Otherwise, the uncovered lines, statements, states in FSM, and signals are considered suspicious. All the suspicious parts constitute the suspicious list. For example, Table 2.7 presents results of coverage and functional analyses for a hardware Trojan (HT-RS232) inserted in the RS232 circuit and looks for a special input sequence 8′ha6–8′h75–8′hc0–8′hff. Upon activation, the hardware Trojan changes the FSM in the transmitter of RS232 from state *Start* to *Stop*, which means RS232 will stop transmitting data (this hardware Trojan is similar to RS232-T700, RS232-T900, and RS232-T901 on Trust-Hub [11]). Table 2.7 indicates that there exist cases that the coverage is not 100% in many cases, and it flags several lines of the RS232 circuit at the register-transfer level as suspicious statements. Further, different test benches present different coverage and by increasing the number of test patterns more coverage can be obtained at the cost of more analysis time.

2.5 Conclusions

In this chapter, the vulnerability of register-transfer-level circuits to hardware Trojans insertion was discussed. Incomplete design specifications and the significant complexity of modern designs create considerable opportunities for rough entities to maliciously manipulate a circuit's functionality. While considerable portion of design development of a circuit is on its verification, traditional verification

Table 2.7 Coverage and functional analyses for hardware Trojan HT-RS232 [10]

Testbench #	Testbench 1	Testbench 2	Testbench 3	Testbench 4	Testbench 5
Test patterns #	2000	10,000	20,000	100,000	1,000,000
Verification time	1 min	6 min	11 min	56 min	10 h
Line coverage (%)	89.5	95.2	98.0	98.7	100
FSM state coverage (%)	87.5	87.5	93.75	93.75	100
FSM transition coverage (%)	86.2	89.65	93.1	96.5	100
Path coverage (%)	77.94	80.8	87.93	97.34	100
Assertion	Successful	Successful	Successful	Successful	Failure

techniques have yet mainly focused on required and specified design requirements. There is a need for a systematic approach to investigate circuit vulnerability to hardware Trojan insertion at the register-transfer level. The vulnerability analysis based on value range is to determine range of existing but unused values that can be used for hardware Trojan insertion. Studying a different vulnerability, the X-analyzer tool identifies dangerous don't cares in a circuit that can be misused for secret keys leakage. Coverage analyses use the existing metrics for circuit verification to catch any possible malicious line in a register-transfer-level circuit. While these techniques have some degrees of success in the vulnerability analysis of circuits at the register-transfer level, their evaluation needs to be extended to industrial sized circuits to evaluate their weaknesses and strengths.

References

1. H. Salmani, M. Tehranipoor, Analyzing circuit vulnerability to hardware Trojan insertion at the behavioral level, in *2013 IEEE International Symposium on Defect and Fault Tolerance in VLSI and Nanotechnology Systems (DFTS)* (2013), pp. 190–195
2. W.H. Harrison, Compiler analysis of the value ranges for variables. IEEE Trans. Softw. Eng. **3**(3), 243–250 (1977)
3. ITC99 benchmarks. http://www.cad.polito.it/downloads/tools/itc99.html. Accessed 22 Jan 2018
4. Simple-as-possible (SAP) microprocessor. http://opencores.org/project,sap. Accessed 22 Jan 2018
5. M. Dale, Verification crisis: managing complexity in SoC designs. https://www.eetimes.com/document.asp?doc_id=1215507. Accessed 22 Jan 2018
6. N. Fern, S. Kulkarni, K.-T.T. Cheng, Hardware Trojans hidden in RTL don't cares – automated insertion and prevention methodologies, in *International Test Conference (ITC)*, October 2015
7. N. Fern, K.-T.T. Cheng, Verification and trust for unspecified IP functionality, in *Hardware IP Security and Trust* (Springer, 2017)

8. N. Fern, I. San, K.-T.T. Cheng, Detecting hardware Trojans in unspecified functionality through solving satisfiability problems, in *Asia and South Pacific Design Automation Conference (ASP-DAC)*, January 2017
9. C. Rebeiro, D. Mukhopadhyay, *High Speed Compact Elliptic Curve Cryptoprocessor for FPGA Platforms* (Springer, Berlin, 2008), pp. 376–388
10. X. Zhang, M. Tehranipoor, Case study: detecting hardware Trojans in third-party digital IP cores, in *2011 IEEE International Symposium on Hardware-Oriented Security and Trust (HOST)* (2011), pp. 67–70
11. Trust-HUB. https://www.trust-hub.org/. Accessed 22 Jan 2018

Chapter 3
Design Techniques for Hardware Trojans Prevention and Detection at the Register-Transfer Level

3.1 Hardware Trojan Prevention at the Register-Transfer Level

3.1.1 Dual Modular Redundant Schedule at High-Level Synthesis

Dual Modular Redundant (DMR) schedule during High-Level Synthesis (HLS) has been proposed to ensure security against hardware Trojans at the register-transfer level [3]. The technique targets hardware Trojans in third-party intellectual properties (3PIPs) that only cause computational output value change (and produces no other impact). Only a third-party vendor is considered untrustworthy, and an adversary or rogue designer only in the third-party house can manipulate the modules/IPs such as decoders, comparators, and multipliers.

Conversion of a circuit behavioral description into its corresponding register-transfer level structure is accomplished through HLS which involves the process of design space exploration (DSE) that includes evaluation of alternative candidate design solutions based on objectives such as area and delay. The conversion process involves multiple subprocesses such as scheduling, allocation, and binding. The process of DSE becomes convoluted with the involvement of an auxiliary variable called loop unrolling factor for control-data flow graph (CDFG) applications as it adds an extra dimension to explore based on conflicting user constraints of area and delay. Loop unrolling plays an important significance in dictating the final area and delay of a design.

Figure 3.1 presents the flow for low-cost security aware high-level synthesis solution for hardware Trojan secured datapath [3]. The aim of proposed approach is to explore the design space of a hardware Trojan Secured DMR schedule comprising of candidate solutions for DMR schedule resource configurations (architecture), candidate loop unrolling factor, and candidate vendor assignment procedure, all hybrid encoded.

H. Salmani, *Trusted Digital Circuits*, https://doi.org/10.1007/978-3-319-79081-7_3

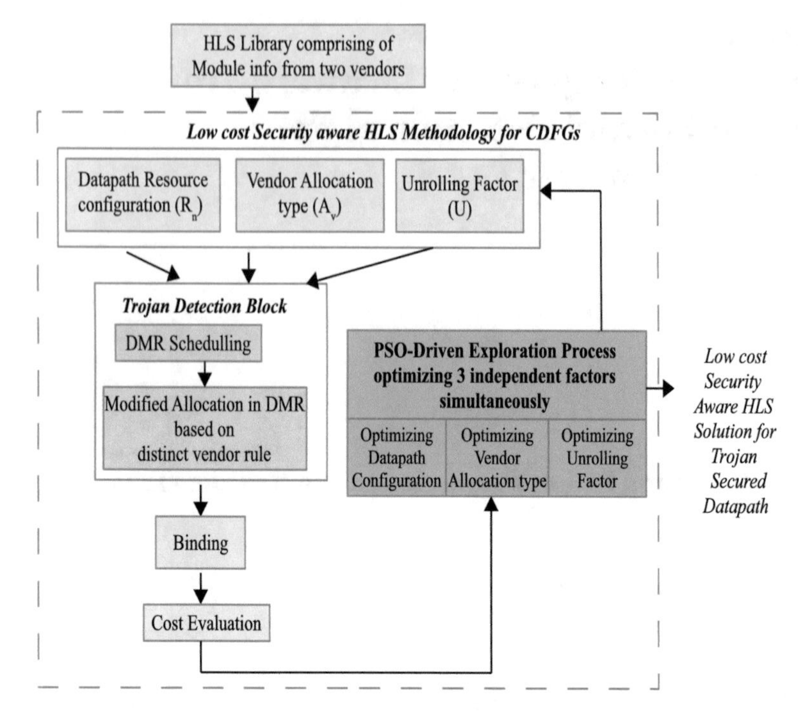

Fig. 3.1 The flow for low-cost security aware high-level synthesis solution for hardware Trojan secured datapath [3]

A solution for Trojan secured DMR schedule is an optimal X_i where $\{X_i\} = \{N(R_1), N(R_2), \ldots, N(R_d), U, P_v\}$, while exploring the design space of a given CDFG and satisfying conflicting user constraints and minimizing the overall cost. X_i indicates a resource set of a particular particle solution with unrolling and allocation procedure information; $N(R_1), N(R_2), \ldots, N(R_d)$ indicates the number of instances of resource type '1', ...,'d'; U indicates the unrolling factor; and P_v indicates vendor allocation procedure type. The problem is formulated as minimizing Hybrid $Cost(A_T^{DMR}, T_E^{DMR})$, for optimal $\{X_i\}$, subjected to $A_T^{DMR} < A_{cons}$ and $T_E^{DMR} < T_{cons}$ and hardware Trojan security. A_T^{DMR} indicates the total area of a DMR design; A_{cons} indicates the user-specified area constraint; T_E^{DMR} indicates the total execution time of a DMR design; and T_{cons} indicates the user specified execution time constraint. The loop unrolling transformation duplicates the body of the loop multiple times controlled by U (The unrolling factor) to expose additional parallelism that may be available across loop iterations. P_v is the vendor allocation procedure capable of holding only a binary value (where P_v = '1' indicates all operations of a specific unit being strictly assigned to resources of the same vendor type, for example, all operations of original unit strictly assigned to the same vendor V_1 and all operations of duplication to the other vendor V_2; while P_v = '0' indicates alternate vendor assignment to operations in a control step of a unit). The variable P_v is crucial for hardware Trojan secured schedule optimization, as both $P_v = $ '0' and $P_v = $ '1' provide vendor distinctness in DMR design resulting in hardware Trojan security.

For hardware Trojan detection in 3PIPs that only change computational output value (and produces no other impact), minimum two distinct third-party vendors are required. Even if the IPs from two different vendors have different timing, but functional similarity of two distinct IPs, allow for comparison at the DMR output. In the proposed approach we only require two distinct vendors for generating a hardware Trojan secured schedule. The technique optimizes the cost of solution by regulating the internal allocation process of two distinct vendors within DMR schedule through a variable P_v during exploration. Imposing a diverse set of 3PIP vendors as security constraints during allocation step for similar operations in DMR design during HLS provides detection of malicious output. The technique simultaneously explores an optimal schedule and optimal U combination for a low-cost hardware Trojan security aware DMR schedule using the Particle Swarm Optimization (PSO) algorithm.

To obtain a hardware Trojan secured DMR schedule, its corresponding DMR schedule is first obtained. The complete duplication is done for all the unrolled operations of the CDFG based on U. Once operations of the CDFG are unrolled, then both the operations of original unit and duplicate unit are concurrently scheduled based on the information of the schedule architecture $(N(R_1), N(R_2), \ldots, N(R_d))$ in the candidate design solution (X_i). This enables to determine the final delay C_T^{DMR}. The DMR logic uses a specific vendor allocation rule to design a hardware Trojan secured schedule. The vendor allocation rule states that two distinct vendors are required for operation assignment in DMR such that the similar operations v of original unit (W^{OG}) and v' of duplicate unit (W^{DP}) are assigned to the distinct vendors. This enables hardware Trojan detection as for cases where no sub-IP exists, it is highly unlikely that different hardware Trojans in different 3PIPs will produce identical wrong outputs. The distinct vendor assignment rule for hardware Trojan detection can be realized in various ways; therefore, the PSO algorithm is used to obtain the most optimal distinct vendor allocation.

3.1.2 Proof-Carrying Hardware

It is possible to express properties of systems in the form of logical statements, and then use theorem provers to prove or disprove the properties of systems expressed as logical statements. The proof-carrying hardware (PCH) framework integrates interactive theorem prover and SAT solvers (shown in Fig. 3.2) to verify security properties of register-transfer level designs [1]. The approach is based on the proof-carrying code (PCC). In PCC, a software customer provides safety properties and untrusted software vendor develops safety proof for these properties. Afterwards, the vendor submits the customer a PCC binary file that includes the formal proof of

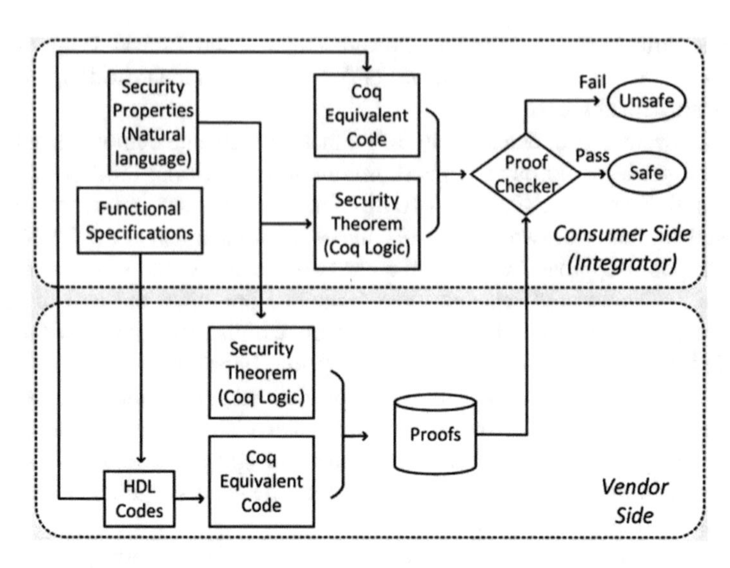

Fig. 3.2 Working procedure of the PCH framework [1]

the safety properties encoded along with the executable code of the software. The customer then validates the PCC binary file using its own proof checker.

Note that the PCH framework does not provide protection against hardware Trojans whose behaviors and attributes are not captured by the set of security properties. Furthermore, there is also an assumption that the attacker has detailed knowledge of the design to identify critical registers and modify them in order to carry out the attack.

In PCH framework, Hoare-logic style reasoning is used to prove the correctness of the RTL code and implementation is carried out using the Coq proof assistant. Coq is indeed a formal proof management system that provides a formal language to write mathematical definitions, executable algorithms, and theorems, as terms in the Gallina specification language, together with an environment for semi-interactive development of machine-checked proofs. In its original implementation, Coq does not recognize hardware description languages and security properties expressed in a natural language. The PCH framework integrates the semantic translation of HDLs and informal security specifications to the Calculus of Inductive Construction (CIC). In the PCH framework, an IP customer provides both functional specifications and a set of security properties to an IP vendor. The functional specifications are materialized into HDL codes. Then the HDL codes and security properties are translated to CIC. Proofs are constructed for security theorems and the transformed HDL code. The HDL code and proof for security properties are combined into a trusted bundle and delivered to the IP customer. The IP customer first generates the formal representation of the design and security properties in CIC. The translated code, combined with formal theorems and proofs, was quickly validated using the proof checker in Coq platform.

3.2 Hardware Trojan Detection at the Register-Transfer Level

3.2.1 *Control-Flow Subgraph Matching*

A verification approach is proposed to detect hardware Trojan at the register-transfer level by exploiting an efficient control-flow subgraph matching algorithm [2]. For a process *P* of a design under verification (DUV) at RTL, a control-flow graph (CFG) for *P* is a tuple $CFG(P) = (B, E, \rho, s, e)$ where $B = b1, \ldots, bn$ is the finite set of basic blocks, that is, sequences of consecutive instructions without any branch; $E \subseteq B \times B$ is the finite set of edges between the blocks such that $(b1, b2) \in E$ if and only if *b2* can be executed after *b1* in at least one of the possible executions of the process *P*; $\rho : E \rightarrow (0, 1]$ is the function such that $\rho(b1, b2)$ is the probability that *b2* follows *b1* during an execution of *P*; $s, e \in B$ are the first and last basic blocks, respectively.

Figure 3.3 presents the overview of the proposed approach. The inputs are the design under verification (DUV) and a hardware Trojan library at the register-transfer level. Three scenarios of hardware Trojans insertion at RTL are considered: (1) an in-house rogue designer intentionally hides malicious behavior in the RTL modules before design verification and synthesis steps; (2) third-party RTL modules provided by untrusted parties have already been manipulated and contain hardware Trojans; and (3) a hardware Trojan might be automatically inserted by a computer-aided design tool used for design syntheses and analyses. Considering the different hardware Trojan insertion scenarios, a hardware Trojan library is developed based

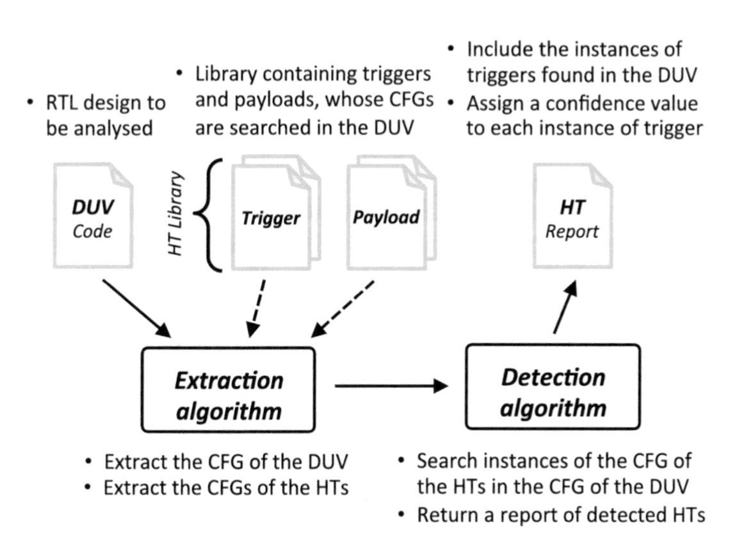

Fig. 3.3 Hardware Trojan detection at the register-transfer level using control-flow subgraph matching [2]

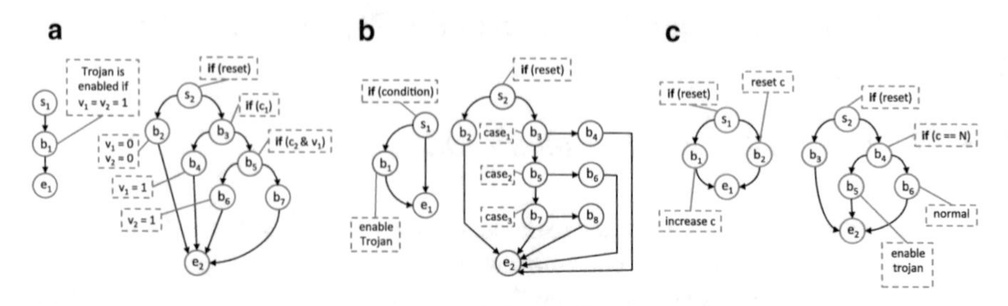

Fig. 3.4 Examples of the CFGs of hardware Trojan triggers in the proposed hardware Trojan library. The dashed boxes represent the instructions (in pseudo-code) inserted in the basic blocks associated to the nodes. In case of a node with two outgoing edges, the left and the right edges correspond to the case in which the branching condition evaluated in the node is true and false, respectively [2]. (**a**) Cheat code. (**b**) Dead machine. (**c**) Ticking time bomb

on known hardware Trojan triggers and their camouflaged variants. The hardware Trojan triggers are classified in three categories: cheat code, dead machines, and time bombs as samples of them are shown in Fig. 3.4.

A *cheat code* is a value (or sequence of values) that enables the payload upon observation. Figure 3.4a presents an example of CFG of a cheat code's trigger based on two processes. A *dead machine* activates the hardware Trojan when specific state-based conditions are met. It can be indeed considered a generalization of cheat code hardware Trojans to realize more complex conditions for hardware Trojan activation. Figure 3.4b shows a possible implementation. A *ticking time bomb* enables the payload when a certain number N of clock cycles has been counted. In fact, if N is sufficiently high, dynamic methods require long simulations, while formal approaches face the state explosion problem. Figure 3.4c illustrates the CFG of a possible implementation of this trigger based on two processes. The hardware Trojan library includes a basic implementation and a configuration file for each of the aforementioned hardware Trojan triggers. The basic implementation is the simplest form of the trigger's code. The configuration file makes the triggers parametric using extension directives and confidence directives. The extension directives are used to automatically modify the CFG of the trigger's basic implementations. This would make it possible for the hardware Trojan detection technique to stay resilient against various implementation of one hardware Trojan trigger. The confidence directives are used to define the structural characteristics of the trigger's CFG. The detection algorithm checks the structural characteristics to determine a confidence value and to avoid false positive, after a possible instance of a HT is found in the DUV.

After establishing the hardware Trojan library, the extraction algorithm is executed. The extraction algorithm creates the CFG for the RTL DUV for each process included in the DUV. The algorithm generates the CFGs of the triggers and payloads included in the HT library as well. In this way, both the structures of the DUV and the HTs are represented with graph-based models that highlight their execution paths and simplify the detection of HTs. In the following, the detection algorithm identifies the HTs by using a subgraph isomorphism algorithm. First, it looks for the parts of the CFG of the DUV that match the CFGs extracted from the triggers of the HT library. Afterward, it analyzes the matched instances to provide confidence values that help discard the false positives. The confidence values take into account the structural characteristics of the CFGs and the related payloads.

Some hardware Trojans can be similar to actual legal code, so it is necessary to give a confidence value for each match returned by the detection algorithm. Four characteristics are considered to determine the confidence value of the match: (c1) presence of variables, (c2) presence of the reset logic, (c3) average distance of the probabilities of the match and the corresponding pattern, and (c4) degree of dependence between the match and the most affine payload. The *c1* verifies if the match uses some variables in the same way of the corresponding pattern. For example, in the case of *time bomb*, it checks if there is a variable used as a counter in the match, but not necessarily with the same name of one in the trigger. The *c2* checks if the match has a reset logic similar to the reset logic of the pattern. For example, in the case of *time bomb*, it checks if a variable (counter) is reset whenever the DUV is reset. The *c3* calculates the distance between the probabilities of traversing each edge in the match and the expected probabilities for traversing the corresponding edges in the pattern; nearer are the probabilities, more is likely that the match is a real instance of the pattern. The *c4* verifies if there are shared variables, that is, registers, between the match and one of the payloads specified in the HT library. To determine final confidence level (β), the four values are linearly combined as $\beta = \alpha_1 \times c_1 + \alpha_2 \times c_2 + \alpha_3 \times c_3 + \alpha_4 \times c_4$ where α_i depends on the trigger and $\sum_i \alpha_i = 1$.

3.3 Conclusions

This chapter reviewed some major works in hardware Trojan prevention and detection at the register-transfer level. Some techniques have suggested the incorporation of various vendors at a fine granularity to prevent hardware Trojan activation. Some have relied on the existence of a hardware Trojan library and searched into a design data/control graph to find subgraphs may match possible hardware Trojans. Embedding security properties into circuit development for later checking has been also recommended with assumption that hardware Trojans behaviors and attributes are known. While the techniques are successful in their goals, their limitations in scalability to large design yet require thorough analyses as they may incur significant area, power, and performance overhead. Further, their evaluation time may considerably increases as designs become larger and more complex. One share assumption among many hardware Trojan detection techniques is the need for a hardware Trojan library that is a challenging issue.

References

1. E. Love, Y. Jin, Y. Makris, Proof-carrying hardware intellectual property: a pathway to trusted module acquisition. IEEE Trans. Inf. Forensics Secur. **7**(1), 25–40 (2012)
2. L. Piccolboni, A. Menon, G. Pravadelli, Efficient control-flow subgraph matching for detecting hardware Trojans in RTL models. ACM Trans. Embed. Comput. Syst. **16**(5s), 137:1–137:19 (2017)
3. A. Sengupta, S. Bhadauria, S.P. Mohanty, TL-HLS: methodology for low cost hardware Trojan security aware scheduling with optimal loop unrolling factor during high level synthesis. IEEE Trans. Comput. Aided Des. Integr. Circuits Syst. **36**(4), 655–668 (2017)

Chapter 4
Circuit Vulnerabilities to Hardware Trojans at the Gate Level

4.1 Circuits at the Gate-Level

In its early stages, a design can be stated using high-level specification languages such as synthesizable subsets of ANSI C/C++/SystemC/MATLAB. The behavioral synthesis analyzes, architecturally constrains, and schedules to create a register-transfer level (RTL) hardware description language (HDL). The RTL circuit then undergoes the logic synthesis to obtain its corresponding gate-level netlist based on a target technology library. A gate-level netlist is simply a description of the connectivity of a circuit. A netlist for a digital circuit contains gates and their interconnections in the simplest form.

The logic synthesis aims at design optimization based on performance, power, or/and area depending on design specifications provided by a costumer. Optimization goals or manipulating design specifications may create undefined or unused zones that provide a malicious designer opportunities to insert hardware Trojan into the gate-level circuit. For example, an optimized state machine may have several unused transitions that can be used to realize a hardware Trojan trigger. Or, an output vector signal in a specific state of a state diagram can be manipulated to covertly leak some key information.

4.2 Analyzing Vulnerabilities Based on Functional Analyses

A static Boolean function analysis, called Functional Analysis for Nearly unused Circuit Identification (FANCI), has been proposed to find signals with weakly affecting inputs [1]. FANCI is grounded on the fact that a hardware Trojan trigger input generally has a weak impact on output signals. FANCI quantifies the dependency between two signals based on a metric called *control value* that measures the degree of control of a signal on another signal in a gate-level netlist. Figure 4.1 presents the identification of vulnerable signals which present small

H. Salmani, *Trusted Digital Circuits*, https://doi.org/10.1007/978-3-319-79081-7_4

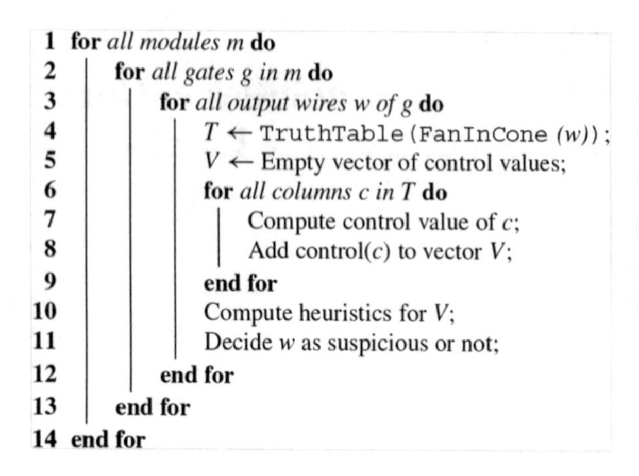

```
1  for all modules m do
2  |  for all gates g in m do
3  |  |  for all output wires w of g do
4  |  |  |  T ← TruthTable(FanInCone(w));
5  |  |  |  V ← Empty vector of control values;
6  |  |  |  for all columns c in T do
7  |  |  |  |  Compute control value of c;
8  |  |  |  |  Add control(c) to vector V;
9  |  |  |  end for
10 |  |  |  Compute heuristics for V;
11 |  |  |  Decide w as suspicious or not;
12 |  |  end for
13 |  end for
14 end for
```

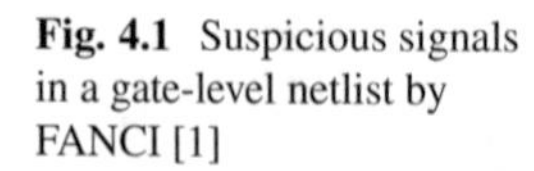

Fig. 4.1 Suspicious signals in a gate-level netlist by FANCI [1]

impacts on signals in their fan-out cone. A functional truth table is constructed for every signal out of signals in their fan-in cone in a gate-level netlist. Given an input signal w_i and an output signal w_o, the control value is the ratio of the number of times the value of w_o changes when w_i changes while other signals in the fan-in cone of w_o are constant (*count*) to the size of the trust table of w_o signal (*size(T)*).

Upon calculating the control values for a target signal and signals in its fan-in cone in a gate-level netlist, FANCI determines a confidence level based on four other measures: *median*, *mean*, *median and mean*, and *triviality*. While mean is the common average and median is the middle value in the list of numbers, triviality calculates a weighted average of the vector of control values. While these measures can identify vulnerable and suspicious signals, they may result in false conclusion due to the circuit implementation. Each measure needs a cutoff threshold to decide a signal is vulnerable or suspicious or not. Such a threshold can be based on a previous knowledge or be determined based on the distribution of computed control values. Figure 4.2 presents results of vulnerability analyses by FANCI over some hared Trojan-free circuits available on Trust-Hub [2]. The results indicate different measures interpret control values differently in most cases for larger circuits.

4.3 Analyzing Vulnerabilities Based on Structural and Parametric Analyses

Functional hardware Trojans are realized by adding or removing gates; therefore, the inclusion of hardware Trojan gates or the elimination of circuit gates affects circuit side-channel signals such as power consumption and delay characteristics, as well as the functionality. To minimize a hardware Trojan's contribution to the circuit side-channel signals, an adversary can exploit hard-to-detect areas (e.g., nets) to implement the hardware Trojan. Hard-to-detect areas are defined as areas

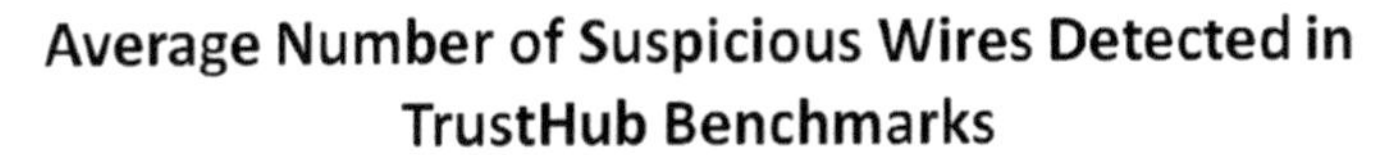

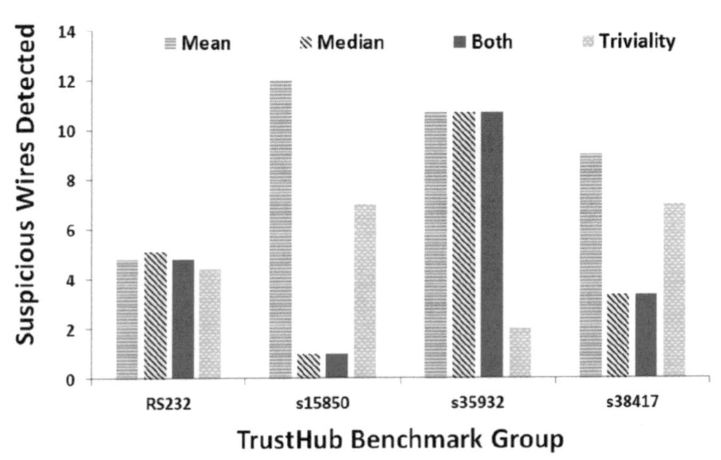

Fig. 4.2 These are the total number of suspicious wires detected by each method for each type of backdoor design on average. For each design and each of the four methods we tried, we always found at least one suspicious wire. Thus, each of the four methods is empirically effective. However, some turned up larger portions of the trigger critical paths, proving to be more thorough for those cases [1]

in a circuit not testable by well-known fault-testing techniques (stuck-at, transition delay, path delay, and bridging faults) or not having noticeable impact on the circuit side-channel signals.

Figure 4.3 shows a vulnerability analysis flow that performs power, delay, and structural analyses on a circuit to extract the hard-to-detect areas [3]. Any transition inside a hardware Trojan circuit increases the overall transient power consumption; therefore, it is expected that hardware Trojan inputs are supplied by nets with low transition probabilities to reduce activity inside the hardware Trojan circuit.

The *Power Analysis* step in Fig. 4.3 is based on analyzing switching activity; it determines the transition probability of every net in the circuit assuming the probability of 0.5 for ‘0’ or ‘1’ at primary inputs and at memory cells’ outputs. Then, nets with transition probabilities below a certain threshold are considered as possible hardware Trojan inputs. The *Delay Analysis* step performs path delay measurement based on gates’ capacitance. This allows to measure the additional delay induced by hardware Trojan by knowing the added capacitance to circuit paths. The Delay Analysis step identifies nets on noncritical paths as they are more susceptible to hardware Trojan insertion and harder to detect their changed delay. To further reduce hardware Trojan impact on circuit delay characteristics, it also reports the paths to which a net belongs to avoid selecting nets belonging to different sections of one path. The *Structural Analysis* step executes the structural transition delay fault testing to find untestable blocked and untestable redundant

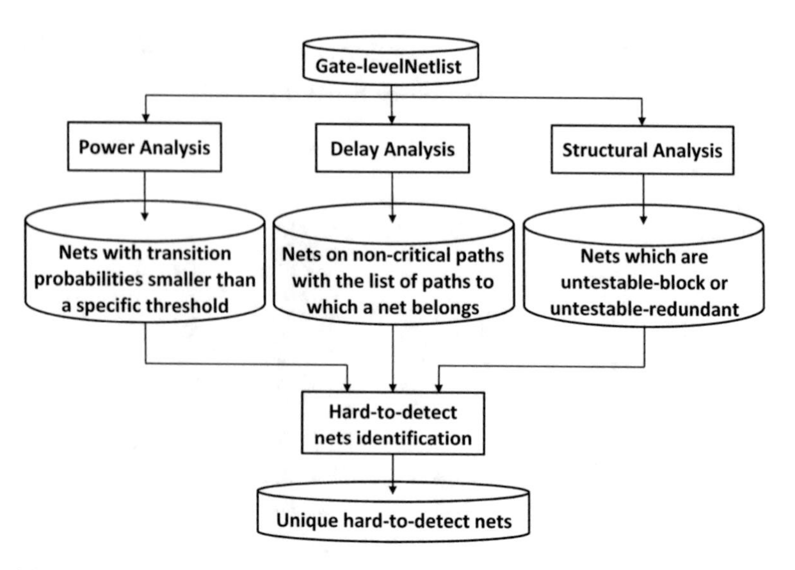

Fig. 4.3 The vulnerability analysis flow [3]

nets. Untestable redundant nets are not testable because they are masked by a redundant logic, and they are not observable through primary output or scan cells. Untestable blocked nets are not controllable or observable by untestable redundant nets. Tapping hardware Trojan inputs to untestable nets hides hardware Trojan impact on delay variations.

At its end, the vulnerability analysis flow reports unique hard-to-detect nets that are the list of untestable nets with low transition probabilities and nets with low transition probabilities on noncritical paths while not sharing any common path. Note that when a hardware Trojan impacts more than one path, it provides greater opportunities for detection. Avoiding shared paths makes a hardware Trojan's contribution to affected paths' delay minimal, which can be masked by process variations, making it difficult to detect and distinguish the added delay from variations. The reported nets are ensured to be untestable by structural test patterns used in production tests. They also have low transition probabilities, so hardware Trojans will negligibly affect circuit power consumption. As the nets are chosen from noncritical paths without any shared segments, it would be extremely difficult to detect hardware Trojans by delay-based techniques.

The vulnerability analysis flow can be implemented using most electronic design automation (EDA) tools, and the complexity of the analysis is linear with respect to the number of nets in the circuit. The flow is applied to the Ethernet MAC 10GE circuit [4], which implements 10 Gbps Ethernet Media Access Control functions. Synthesized at 90 nm Synopsys technology node, the Ethernet MAC 10GE circuit consists of 102,047 components, including 21,830 flip-flops. The Power Analysis shows that out of 102,669 nets in the circuit, 23,783 of them have a transition probability smaller than 0.1, 7003 of them smaller than 0.01, 367 of them smaller than 0.001, and 99 of them smaller than 0.0001. The Delay Analysis indicates that the largest capacitance along a path, representing path delay, in the circuit is 0.065717825 pF, and there are 14,927 paths in the circuit whose path capacitance is smaller than 70% of the largest capacitance, assuming that paths longer than 70% in a circuit can be tested using testers. The Structural Analysis finds that there is no untestable fault in the circuit. By excluding nets sharing different segments of one path, there are 494 nets in the Ethernet MAC 10GE circuit considered to be areas where hardware Trojan inputs could be used while ensuring the high difficulty of detection based on side-channel and functional test techniques.

4.3.1 Hardware Trojan Ranking

A hardware Trojan's impact on circuit characteristics depends on its implementation. Hardware Trojan inputs tapped into nets with higher transition probabilities will aggrandize switching activity inside the hardware Trojan circuit and increase its contribution to circuit power consumption. Furthermore, the hardware Trojan might affect circuit delay characteristics due to additional capacitance induced by extra routing and hardware Trojan gates. Hardware Trojan detectability can establish a fair comparison among different hardware Trojan detection techniques since it is based on induced variations by a hardware Trojan in side-channel signals.

The hardware Trojan detectability metric is determined by (1) the number of transitions in the hardware Trojan circuit and (2) extra capacitance induced by hardware Trojan gates and their routing. This metric is designed to be forward-compatible with new approaches for hardware Trojan detection by introducing a new variable, for example, a quantity related to the electromagnetic field.

Transitions in a hardware Trojan circuit reflect hardware Trojan contribution to circuit power consumption, and hardware Trojan impact on circuit delay characteristic is represented by measuring the added capacitance by the hardware Trojan. Assuming A_{Trojan} represents the number of transitions in the hardware Trojan circuit, S_{Trojan} the hardware Trojan circuit size in terms of the number of cells, A_{TjFree} the number of transitions in the hardware Trojan-free circuit, S_{TjFree} the hardware Trojan-free circuit size in terms of the number of cells, TIC the added capacitance by hardware Trojan as hardware Trojan-induced capacitance, and C_{TjFree} the hardware Trojan-affected path with the largest capacitance in the corresponding hardware Trojan-free circuit, hardware Trojan detectability ($T_{Detectability}$) at the gate level is defined as

$$T_{Detectability} = |t| \tag{4.1}$$

where

$$t = \left(\frac{A_{Trojan}/S_{Trojan}}{A_{TjFree}/S_{TjFree}}, \frac{TIC}{C_{TjFree}} \right) \tag{4.2}$$

$T_{Detectability}$ at the gate level is calculated as follows:

1. Apply random inputs to a hardware Trojan-free circuit and obtain the number of transitions in the circuit (A_{TjFree}).
2. Apply the same random vectors to the circuit with a hardware Trojan and obtain the number of transitions in the hardware Trojan circuit (A_{Trojan}).
3. Perform the Delay analysis on the hardware Trojan-free and hardware Trojan-inserted circuits.
4. Obtain the list of paths whose capacitance is changed by the hardware Trojan.
5. Determine the hardware Trojan-affected path with the largest capacitance in the corresponding hardware Trojan-free (C_{TjFree}) and the added capacitance (TIC).
6. Form the vector t (4.2) and compute $T_{Detectability}$ as defined in Eq. (4.1). Note that hardware Trojan detectability represents the difficulty of detecting a hardware Trojan.

Tables 4.1 and 4.2 show detailed analysis of a selected number of gate-level benchmarks. In Table 4.2, Column 3 indicates that b19 circuit, in Row 2, is the largest circuit in size (62835) among the selected circuits. Table 4.1 also shows the number of nets with transition probability less than 0.0001 in b19, 4530

Table 4.1 Design vulnerability analysis of a selected number of hardware Trojan-free circuits presented at [2]

	Power analysis					Delay analysis		Structural analysis
Circuit	# Nets	<0.1	<0.01	<0.001	<0.0001	Critical path capacitance (pF)	<70% of critical path capacitance	# untestable faults
b19	70,259	14,482	8389	5533	4530	0.37723719	474,358	8
s38417	5669	589	291	219	69	0.050146392	41,901	0
s38584	7203	817	197	85	30	0.044666893	27,689	0
s35932	6269	0	0	0	0	0.00851317	3156	0

Table 4.2 The detectability ($T_{Detectability}$) of a selected number of gate-level hardware Trojans inserted in the circuits in Table 4.1

Trojan	A_{TjFree}	S_{TjFree}	A_{Trojan}	S_{Trojan}	$TIC(pF)$	$C_{TjFree}(pF)$	$T_{Detectability}$
b19-T100	4,037,383	62,835	0	83	0.000945429	0.037849663	0.024978531
s38417-T100	2,717,682	5329	59	11	0.004167929	0.032341219	0.139390942
s38417-T200	2,717,682	5329	1328	11	0.005313744	0.030518107	0.4108472958
s38417-T300	2,717,682	5329	257	15	0.000457899	0.03078216	0.0484715731
s38584-T100	423,986	6473	705	9	0.000945429	0.015039353	1.2587797351
s38584-T200	423,986	6473	0	83	0.000414827	0.029841803	0.0139008691
s38584-T300	423,986	6473	16	731	0.012461155	0.004379333	2.8457800321
s35932-T100	353,304	5426	354	15	0.000494403	0.006985635	0.8664403555
s35932-T200	353,304	5426	733	12	0.003160179	0.009732743	1.2628060457
s35932-T300	353,304	5426	738	36	0.000497869	0.008230419	0.3753278457

in Row 2 and Column 6, is larger than that of the other circuits, and b19 has considerable number of paths whose capacitances are less than 70% of its critical path' capacitance, 474358 in Column 8. Further, there are eight untestable faults in b19, in Column 9. These provide significant opportunity for implanting hardware Trojans resilient against power and delay side-channel analyses in b19. Table 4.2 confirms that b19-T100 with $T_{Detectability} = 0.024978531$, in Column 8, is the second most difficult hardware Trojan to detect as no transition inside the hardware Trojan is observed, 0 in Column 4, and it induces small capacitance, 0.000945429 pF in Column 6, on a noncritical path, 0.037849663 pF in Column 7. s38584-T200, in Row 7, has the lowest detectability, 0.0139008691 in Column 8; similar to b19-T100, there is no switching activity in s38584-T200, 0 in Column 4, and s38584-T200 induces less capacitance, 0.000414827 pF in Column 6, on a shorter path, 0.029841803 pF in Column 7, compared to b19-T100.

4.4 Analyzing Vulnerabilities in Finite State Machines and Design-for-Test Structures

Beyond pure gates and paths analyses, it is possible to perform functional-oriented vulnerability analyses at the gate level [5, 6]. A finite state machine (FSM) is the heart of any computing system as it controls all operations in the system. An FSM mainly consists of some memory elements that keep the current state of system and a combinational circuit that determines the next state of system. During logic synthesis, don't-care states and transitions might be introduced that are not intended at the register transfer level. Traditional verification techniques may not be capable of identifying and analyzing introduced don't cares; therefore, a rogue designer may exploit them to realize stealthy hardware Trojans with small footprints [5]. Using design-for-test (DFT) architectures such as the scan chain architecture is a common practice considering today's sophisticated designs with multimillion transistors and a dozen functionality. Such architectures facilitate testing of a manufactured circuit by providing extensive controllability and observability of internal signals through a limited number of the circuit's primary outputs. However, from security perspective, DFT architectures can be exploited to access key information inside a circuit [6].

Nahiyan et al. [5] presents a framework that accepts a gate-level netlist along with FSM synthesis report and RTL FSM report. Using FSM synthesis report and RTL FSM report, Don't-care states and transitions are extracted. If a don't-care state exists that has direct access to a protected state, such a don't-care state introduces a vulnerability in the FSM as the attacker becomes enabled to utilize the don't-care state to insert a hardware Trojan to gain access to the protected state. Knowing the set of protected states provided by a designer, the framework isolates dangerous don't-care states (DDCS) and uses the following metric to evaluate the vulnerability of an FSM to hardware Trojan insertion:

$$VF_{Tro} = \frac{\text{Total number of } s'}{\text{Total}_{Transition}} \quad \text{where} \quad s' \in DDCS \tag{4.3}$$

Table 4.3 Vulnerability analysis for scheme I and scheme II of RSA [5]

	VF_{Tro}
Scheme I	0
Scheme II	0.1

The AVFSM framework is to a simplified FSM of an RSA encryption module implementing the Montgomery ladder algorithm [7]. This FSM is composed of seven states, {Idle, Init, Load1, Load2, Multiply, Square, Result}. Here, the attacker's objective is to bypass the intermediate rounds of 'Square' and 'Multiply' states and access the 'Result' state to get either the key or premature result of RSA encryption. Therefore, for this FSM, the set of protected states is P = {Result} and set of authorized states is L = {Square}. Two encoding schemes {000, 001, 010, 011, 100, 101, 110} and {001, 010, 011, 100, 101, 110, 000} represented as scheme I and scheme II, respectively to implement the FSM. The result reported by AVFSM framework is shown in Table 4.3.

The scan chain architecture provides access to internal flip-flops in a design. While it significantly enhances design testability, it may compromise design confidentiality by leaking key information to the outside of chip. Test points are added to eliminate hard-to-test areas. Like the scan architecture, it may facilitate various attacks on a chip. The Joint Test Action Group (JTAG) improves testability of SoC design by providing access to primary inputs and outputs of comprising IPs. However, JTAG can extend the attack surface by providing access to individual module inside the chip [6]. Violation of integrity at the gate-level netlist is defined as the existence of some inputs or control points that can be used for control of a critical module and drive it to some unspecified/dangerous states [6].

The integrity analysis flow [6] identifies control point(s) of an interested asset. The control points are considered as a set of activation vectors in the fan-in of the asset to force it to change its value or enable or disable some features. An integrity report is generated at the end of the analysis. The report contains the list of flagged vulnerabilities and corresponding activation vectors for any asset.

As a case study, the integrity analysis flow is applied to the controller circuit of a MIPS microprocessor [6]. The MIPS microprocessor is a pipeline architecture with five stages including the instruction decode stage. The control path transforms a given opcode to individual instruction control lines needed by proceeding stages. The control path generates control bits, such as MemRead, MemWrite, PCWrite, for the multiplexers, the data memory, and the ALU control. These control bits are expected to be only controlled by the opcode and should not be influenced by user controlled inputs. Thus, control signals MemRead, MemWrite, and PCWrite are analyzed for integrity. Table 4.4 shows the results of their integrity vulnerabilities. Column 3 shows the number of primary inputs and SFFs that an adversary needs to control the asset. An adversary can exploit these vulnerabilities to gain access to memory, escalate user privileges, or modify the program counter value to change the normal flow of a program.

Table 4.4 Integrity analysis of MIPS memory access [6]

Asset	Function	Detected control points
MemRead	Read from memory	5
MemWrite	Write to memory	5
PCWrite	Change program counter	5

4.5 Conclusions

The hardware trust community needs to be placed in a common ground to more effectively address the hardware Trojan detection problem. Efforts have be paid to develop automatic vulnerability analysis tools to determine roots of vulnerabilities in gate-level netlist and quantify their severity. Some work studies the vulnerabilities in terms of gates and timing paths, and some others consider functionality at the module level to analyze susceptibility of designs to malicious modifications. The vulnerability analyses help circuit designers to study the security perspective of their designs and ensure that used design techniques for optimization or testability do not create any room for hardware Trojans. Further, deep vulnerability analyses assist in developing effective countermeasure to prevent and detect hardware Trojans.

References

1. A. Waksman, M. Suozzo, S. Sethumadhavan, FANCI: identification of stealthy malicious logic using boolean functional analysis, in *Proceedings of the 2013 ACM SIGSAC Conference on Computer and Communications Security* (2013), pp. 697–708
2. Trust-HUB. https://www.trust-hub.org/. Accessed 22 Jan 2018
3. H. Salmani, M. Tehranipoor, R. Karri, On design vulnerability analysis and trust benchmarks development, in *2013 IEEE 31st International Conference on Computer Design, ICCD 2013*, Asheville, NC, 6–9 October 2013, pp. 471–474
4. Ethernet 10GE MAC. http://opencores.org/project,xge_mac. Accessed 22 Jan 2018
5. A. Nahiyan, K. Xiao, K. Yang, Y. Jin, D. Forte, M. Tehranipoor, AVFSM: a framework for identifying and mitigating vulnerabilities in FSMS, in *2016 53nd ACM/EDAC/IEEE Design Automation Conference (DAC)* (2016), pp. 1–6
6. G.K. Contreras, A. Nahiyan, S. Bhunia, D. Forte, M. Tehranipoor, Security vulnerability analysis of design-for-test exploits for asset protection in SoCs, in *2017 22nd Asia and South Pacific Design Automation Conference (ASP-DAC)* (2017), pp. 617–622
7. B. Sunar, G. Gaubatz, E. Savas, Sequential circuit design for embedded cryptographic applications resilient to adversarial faults. IEEE Trans. Comput. **57**(1), 126–138 (2008)

Chapter 5
Design Techniques for Hardware Trojans Prevention and Detection at the Gate Level

5.1 Hardware Trojan Prevention at the Gate Level

5.1.1 Information Flow Tracking for Hardware Trojan Prevention

A hardware Trojan may aim to compromise integrity and confidentiality of design. Gate-Level Information Flow Tracking (GLIFT) has been proposed to precisely measure and manage all digital information flows in the underlying hardware [3]. Data are assigned a label to indicate the level of their security in information flow tracking (IFT). The label is then propagated along with data through the system. IFT monitors the flow of information to check whether secret data pass to an unclassified domain or high-integrity data are violated by an untrusted party. For example, Fig. 5.1a presents a 2-input AND gate, where A, B, and O are the inputs and output of AND-2, and a_t, b_t, and o_t are the security labels of A, B, and O, respectively. The security labels can be either '1' or '0'. It can be assumed '1' indicated untrusted, so A is trusted input ($a_t = 0$) and B is untrusted ($b_t = 1$) in Fig. 5.1b. Figure 5.1b shows whether it is possible to trust the out O by analyzing a_t.

It is possible that security labels take more than two values in practice. To address this requirement, GLIFT constructs a library of 2-input primitive gates with m security labels. For a given digital circuit, it is synthesized to a gate-level netlist composed of the primitive gates found in the GLIFT library. Then, one can discretely instantiate tracking logic for each primitive gate in the netlist through a constant-time mapping operation. Figure 5.2 illustrates such a constructive approach [3].

To calculate the security level of a signal, GLIFT defines laws similar to Boolean logic identities such as *Absorption*, *Distributive*, *Associative*, and *Dot Product*. For example, for 2-input AND gate with A and B as the inputs and O as the output and with multilevel security for each input and output (A_t, B_t, and O_t), $O_t = A_t \bigodot B_t$ when $A = B = 0$, and $O_t = A_t \bigoplus B_t$ when $A = B = 1$ where $\bigoplus$ and $\bigodot$ are defined as the least upper and greatest lower bound operators, respectively. Applying GLIFT

H. Salmani, *Trusted Digital Circuits*, https://doi.org/10.1007/978-3-319-79081-7_5

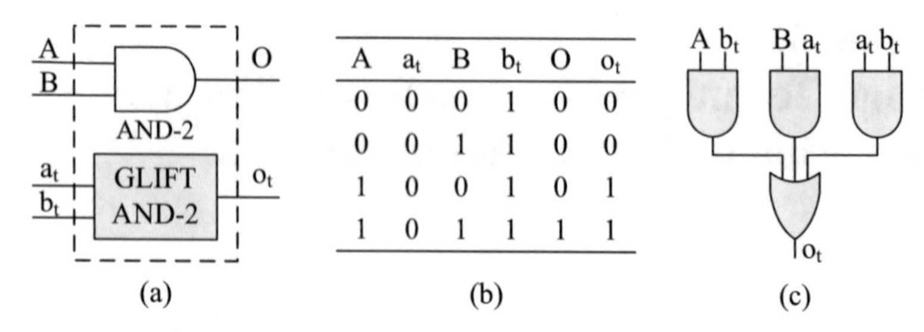

A	a_t	B	b_t	O	o_t
0	0	0	1	0	0
0	0	1	1	0	0
1	0	0	1	0	1
1	0	1	1	1	1

Fig. 5.1 The GLIFT method. (**a**) I/O of AND-2 and its GLIFT logic; (**b**) partial truth table of GLIFT logic for AND-2; (**c**) GLIFT logic of AND-2 [3]

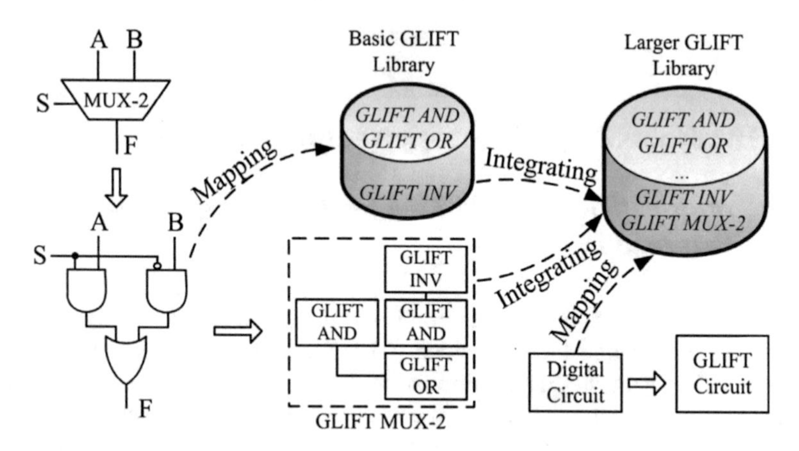

Fig. 5.2 The constructive method for GLIFT logic generation [3]

on an optimized circuit may result as false positive (i.e., incorrect conclusions on integrity compromise or information leakage). To eliminate false positives, GLIFT is being applied on a circuit without static hazard (i.e., any function contains all its prime implicants). However, finding all prime implicants has been proven an NP-hard problem. To eliminate all logic hazards while retaining acceptable computational complexity, synthesis of static-logic-hazard-free circuits using binary decision diagrams (BDDs) is suggested.

GLIFT can be used as a design technique to prevent leakage of secret information by a hardware Trojan at the gate level by tracking the flow information between gates in a circuit. Meanwhile, it should be considered that GLIFT incurs overhead. On average, GLIFT logic created using the BDD method reports 4.51× in total more number of operators, 3.65× in area, 1.74× in delay, and 3.43× in power, respectively, than those generated by the constructive method which includes all prime implicants of a logic function.

5.2 Hardware Trojan Detection at the Gate Level

5.2.1 *Signal Correlation-Based Clustering for Hardware Trojan Detection*

A simulation-based technique for detecting hardware Trojans using a statistical correlation clustering is proposed in [1]. The technique postulates a hardware Trojan logic has weak statistical correlation with the rest of the circuit. The technique consists of three major steps; in the first step, a correlation-based similarity weight is computed for the input-output pairs of each gate in a circuit using the simulation data. In the second step, the gate-level design is converted to a circuit graph, and edges are weighed by the similarity values calculated in the previous step. It is expected that hardware Trojan triggers are driven by gates with very low controllability, and hardware Trojan payloads are appeared as gates with a low observability. Therefore, it is likely that a correlation-based weighting would identify such gates as outliers of the graph. To detect that, in the third step, a density-based clustering algorithm is applied to flag the outliers.

To generate simulation data, functional tests are generated and these test patterns target different regions of the circuit. It is aimed to excite as many nodes as possible to estimate the statistical correlation between adjacent nodes in the circuit graph. The output of a gate undergoes switching activities when the inputs of the gate change their values. Any input that has statistically higher correlation to the output indeed determines the value of the output. Considering signals as nodes of the circuit graph and gates connecting signals as its edges, the correlation between two adjacent signals, where one signal is input of a gate and the other signal is the output of the same gate, determines the weight of the edge.

To determine the weight of an edge, the cross-correlation of all adjacent signal pairs is calculated and then the energy of every resulted signal out of the cross-correlation is calculated. The calculated energy can show and quantify how much two signals are similar to each other. To perform the outlier analysis over the graph, the structural similarity between any adjacent signals based on the energy between them is calculated as

$$\sigma(u, v) = \frac{\sum_{x \in \Gamma(u) \cap \Gamma(v)} w(u, x).w(v, x)}{\sqrt{\sum_{x \in \Gamma(u)} w^2(u, x)}\sqrt{\sum_{x \in \Gamma(v)} w^2(v, x)}} \tag{5.1}$$

where $u \in V$ and $v \in V$ are adjacent vertices connected with the edge $e \in E$ and weight $w(u, v) \in W$ in the weighted graph $G = (V, E, W)$. The weight $w(u, u)$ is defined as 1. $\Gamma(u)$ denotes the neighborhood of the node u including u and its all adjacent vertices, and formally stated as $\Gamma(u) = \{v \in V | \{u, v\} \in E\} \cup \{u\}$.

After obtaining σ for every edge between any two nodes u and v, the distance between u and v is calculated as the sum of the inverse of σ of all edges on the

(a)

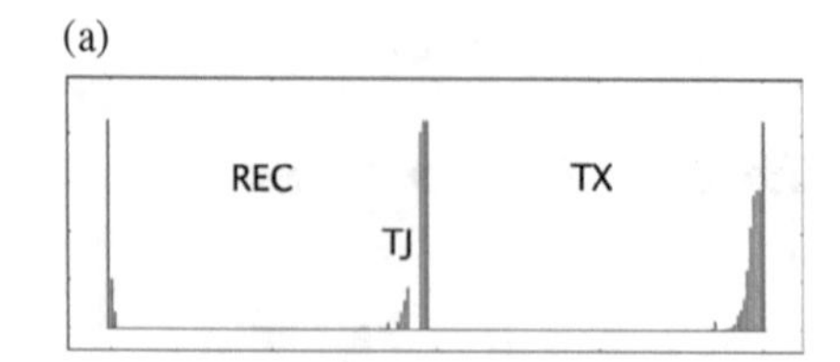

(b)

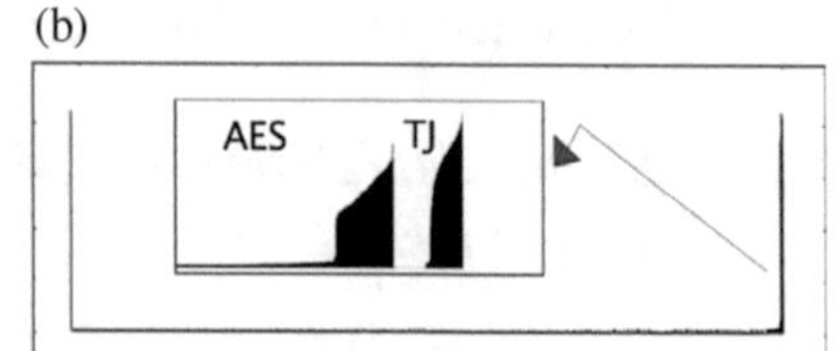

Fig. 5.3 Types of reachability plots observed with TrustHub Trojan benchmarks [1]. (**a**) Reachability plot for RS232-800 showing the receiver (REC) and the transmitter (TX) modules of the UART circuit with hardware Trojan (TJ) logic pushed to the border of the REC cluster. (**b**) Reachability plot for AES-1800 with the hardware Trojan (TJ) logic appearing as a separate cluster at the end of the ordered list

Table 5.1 The results of signal correlation-based clustering for hardware Trojan detection in selected Trust-HUB benchmarks [1]

Name	TPR (%)	SPC (%)
s15850-100	61	99
s35932-200	27	99
s38417-100	100	99
s38584-200	99	98
AES-1800	92	99
wb-conmax-200	28	96
PIC16F84-100	75	96
RS232-800	80	94

shortest path between u and v. Then the Ordering Points To Identify the Clustering Structure (OPTICS) that is a density-based clustering algorithm is applied to obtain the reachability plot.

It is expected that a hardware Trojan delivers functionality that is different from the rest of circuit. Divergence in functionality manifests itself as an outlier with high reachability-distances to the borders of the clusters or explicitly showing the malicious logic as a different cluster in case of a large hardware Trojan.

For example, Fig. 5.3 shows the result of applying the signal correlation-based clustering technique on RS232-800 and AES-1800 available on Trust-Hub [8]. The results show that inserted hardware Trojans are distinguishable and stay far from the main circuits. Table 5.1 shows the true positive rate (TPR) and the specificity (SPC) for various trust benchmarks in Trust-Hub [8]. The average TPR is about 70% while TPR ranges from 27% to 100%. The TPR results show the percentage of hardware Trojan gates that are correctly being identified. The SPC results tell the ratio of the true negatives over the number of non-hardware-Trojan gates. The SPC is the fraction of gates that are falsely flagged as being suspicious and will need to be ruled out in detailed review. The results show the average SPC is 97.5%, and the worse case value of SPC is 94% for RS232-800.

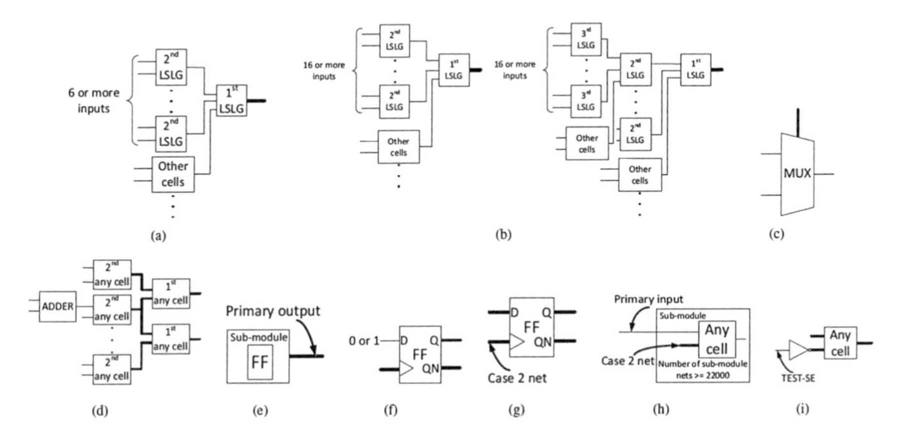

Fig. 5.4 Weak classification nets [4]. (**a**) Case 1. (**b**) Case 2. (**c**) Case 3. (**d**) Case 4. (**e**) Case 5. (**f**) Case 6. (**g**) Case 7. (**h**) Case 8. (**i**) Case 9

5.2.2 Score-Based Classification for Hardware Trojans Detection

A score-based classification method is presented for identifying hardware Trojans in [4]. The technique first extracts features for hardware Trojan nets based on hardware Trojans introduced at Trust-Hub [8]. Figure 5.4 depicts the extracted features of hardware Trojan nets. Nets with any of these features are called the weak classification nets. In Fig. 5.4, bold signals indicate weak classification nets, and a low switching logic gate (LSLG) refers to either AND, NAND, OR, or NOR gate.

While a normal net can take one of the cases in Fig. 5.4. Considering the fact, weak classification nets are divided into groups. Group 1 consists of Case 1, Case 2, Case 3, Case 4, and Case 5, and Group 2 consists of Case 6, Case 7, Case 8, and Case 9. Nets in Group 1 are scored 1, and nets in Group 2 are scored 2. A net may belong to several cases in Fig. 5.4. As a result, the score of a net is the sum of all scores associated with the cases to which the net belongs. It is expected that a hardware Trojan does not frequently interfere circuit functionality; it therefore remains in constantly dormant state (e.g., a hardware Trojan net stays at '0'). To further distinguish hardware Trojan nets from the others, random input vectors are applied to primary inputs of a circuit under test for 1M clock cycles, and all the max score nets obtained in the previous subsection are monitored. It is possible that a normal net even with low score remains constant for a very long time, and it is wrongly identified as a hardware Trojan net. It has been observed that in this scoring, the number of max score nets will decrease if the max score is higher and it will increase if the max score is lower. Therefore, gate-level netlists are expected to be HT-free if the number of max score nets is large enough.

Based on above observations, it is assumed that a net is a hardware Trojan net if it satisfies one of the following conditions: (1) if the score of net is three or more,

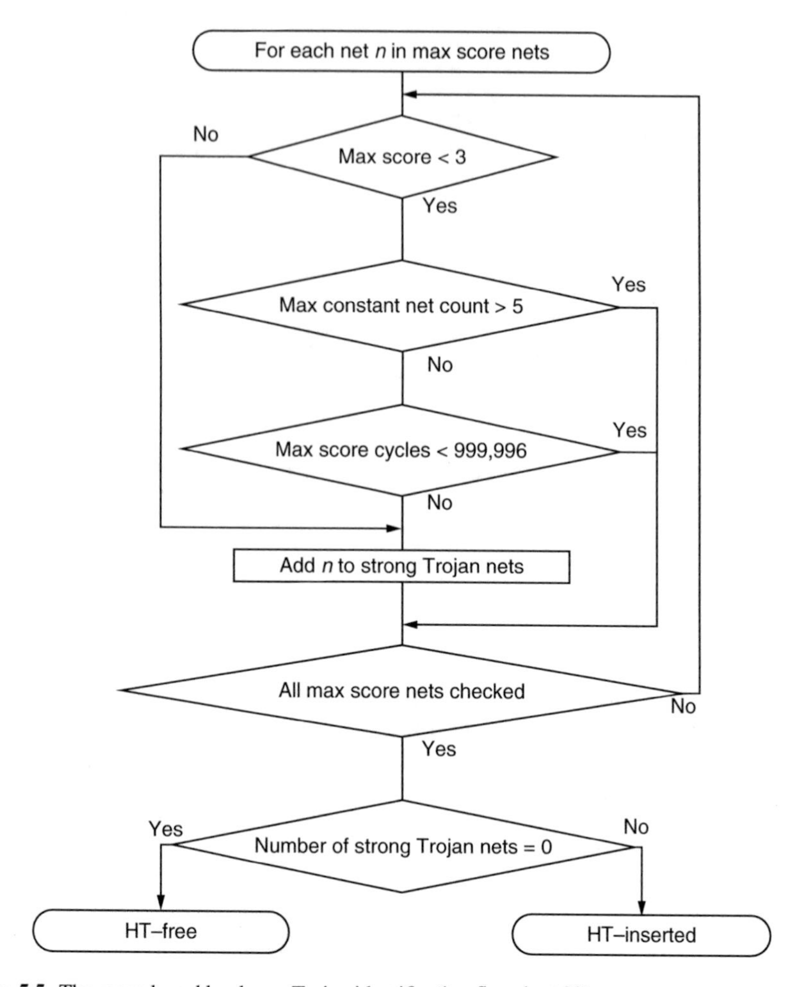

Fig. 5.5 The score-based hardware Trojan identification flowchart [4]

net is a hardware Trojan net; (2) if maximum constant cycles of net are 999,996 or more and the number of the maximum score nets in the netlist is five or less, the net is a hardware Trojan net. Figure 5.5 presents the score-based hardware Trojan identification flowchart.

Figure 5.6 shows 3D illustration of three hardware Trojan factors for max score nets in all the gate-level Trust-Hub benchmarks. Since hardware Trojan nets in Trust-Hub are known, X shows a hardware Trojan net and O shows a normal net, that is, X and O are the correct answers. Shaded rectangles in Fig. 5.6 demonstrate results of the score-based hardware Trojan identification technique. Shaded rectangles only include the correct hardware Trojan nets marked by 'X', and at least one hardware Trojan net in each benchmark is detected by the technique.

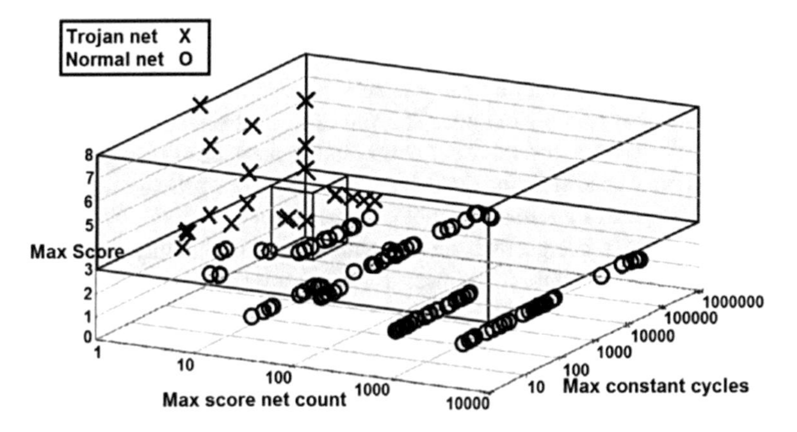

Fig. 5.6 3D illustration of three hardware Trojan factors [4]

5.2.3 The Controllability and Observability Hardware Trojan Detection (COTD)

To reduce hardware Trojan detectability, the hardware Trojan trigger might be connected to nets with low activity to create a rare triggering vector, and the hardware Trojan payload might restitch nets whose deviations are not readily observable. As a result, it is expected that hardware Trojan signals present a very low switching activity and small correlation with other signals in a circuit as considered in [1, 4] and [9]. However, the circuit may contain some genuine signals with low switching activity, and they can be wrongly identified as hardware Trojan signals (i.e., false positive). On the other hand, a smart implementation of hardware Trojan can realize a low-active hardware Trojan trigger whose hardware Trojan trigger does not match any cases studied in [4]. It is also possible that a hardware Trojan has no trigger as it is an always-on hardware Trojan, and its payload is not connected to any net in the circuit. For example, an inverter-based ring-oscillator hardware Trojan only consists of an odd number of inverters. This hardware Trojan is not connected to any internal net and is to reduce design reliability over the time. Such a hardware Trojan can easily evade the hardware Trojan detection techniques presented in [1] and [4] (i.e., false negative).

Testability is a relative measure of the effort or cost of testing a logic circuit, and it can be used to identify nets with poor testability [10]. Testability considers the circuit implementation and gates' interconnections. Low-testability signals more infrequently determine circuit primary outputs and its states; therefore, their manipulation highly remains hidden. Testability analysis can be performed by calculating the controllability and observability of each signal. The Sandia Controllability/Observability Analysis Program (SCOAP) is the most popular testability program that measures testability of each signal s in a circuit logic based on several numerical values including $CC0(s)$—combinational 0-controllability of s, $CC1(s)$—combinational 1-controllability of s, and $CO(s)$—combinational observability of s. These combinational testability measures roughly determine the number of signals that must be manipulated in order to control or observe s from primary inputs or at primary outputs. The values of controllability measures range between 1 and ∞, while the values of observability measures range between 0 and ∞. As a boundary condition, the $CC0$ and $CC1$ values of a primary input are set to 1, and the CO value of a primary output is set to 0 [10].

From the security perspective, signals with low controllability or low observability are more susceptible to be used for the hardware Trojan trigger and the hardware Trojan payload. While a target signal might be highly observable but hardly controllable or vice versa, the pair $\langle CC, CO \rangle$ is created for each signal in the circuit and the magnitude of pair is defined as

$$|\langle CC, CO \rangle| = \sqrt{CC^2(s) + CO^2(s)}. \tag{5.2}$$

While the value of CO can be directly obtained from the SCOAP program, the CC value is defined as

$$CC(s) = \sqrt{CC0^2(s) + CC1^2(s)}. \tag{5.3}$$

It is expected for hardware Trojan signals to have low testability to avoid frequent impact on design functionality. Figure 5.7 presents the controllability and observability hardware Trojan Detection (COTD) flow [5]. It takes a gate-level netlist as the input, and the *Controllability and Observability Analyses* step performs the controllability and observability analyses to determine controllability and observability values, that is, $CC0$, $CC1$, and CO values. Afterward, the *Unsupervised Clustering Analysis* step is performed to cluster signals based on their controllability and observability values. Two signal lists are produced: *Trojan Signals List* and *Genuine Signals List*. The Genuine Signals List only contains all genuine/original signals in the netlist. The Trojan Signal List only contains all hardware Trojan signals if any hardware Trojan exists. If this list is empty, the circuit is hardware Trojan free; otherwise, the circuit is hardware Trojan inserted.

One major issue associated with hardware Trojan detection is the lack of golden circuit as a reference. Unsupervised learning is a type of machine learning algorithm used to explore datasets consisting of input data without labeled responses. Cluster analysis or clustering is the most common unsupervised learning method used for grouping data. It groups a set of objects in such a way that objects in the same group/cluster are more similar to each other than to those in other groups/clusters. The clusters are modeled using a measure of similarity metrics such as Euclidean. Two factors determine the quality of clustering: intra-cluster distance and intercluster distance. While clustering, it is desired to maximize intercluster distances and to minimize intra-cluster distances. One of the most common clustering algorithms is k-means clustering that strives to meet both goals at the same time [6].

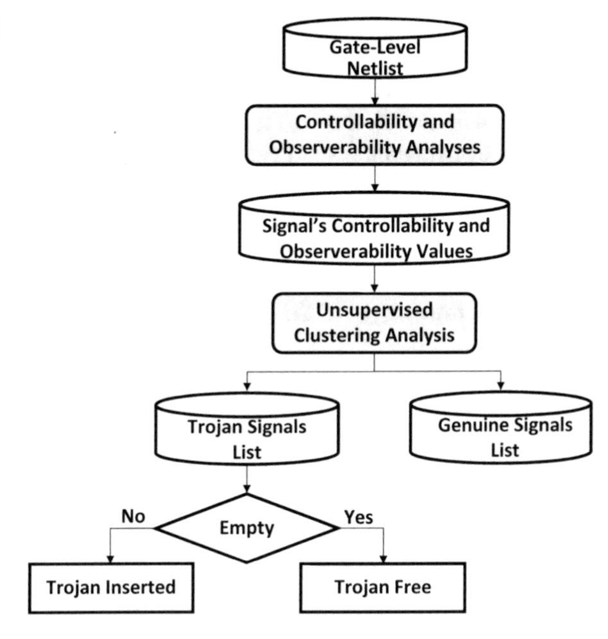

Fig. 5.7 The COTD flow [5]

k-means clustering aims to partition n observations into k clusters in which each observation belongs to the cluster with the nearest mean, serving as the center (or centroid) of the cluster. Given a set of observations $(x_1, x_2, \ldots, x_n)$, where each observation is a d-dimensional real vector, k-means clustering aims to partition the n observations into k sets $S = \{S_1, S_2, \ldots, S_k\}$ so as to minimize the within-cluster sum of squares. In other words, its objective is to find:

$$arg\ min_S \sum_{i=1}^{k} \sum_{x \in S_i} ||x - \mu_i||^2 \tag{5.4}$$

Using Eq. (5.2), the COTD technique obtains the magnitude of pair $\langle CC, CO \rangle$ for every signal in a circuit. Then it passes them in the form of a 2-dimensional dataset to the k-means algorithm where $k = 3$. Signals can be divided into three clusters ($k = 3$): (1) genuine signals whose both CC and CO values are small, and hardware Trojan signals (2) with large CC values or (3) with large CO values. Therefore, the k parameter is set to 3 to distinguish hardware Trojan signals from genuine signals. It is possible for a hardware Trojan signal to have both large CC and CO values. Such a hardware Trojan signal is still distinguished and reported in *Trojan Signal List*.

As hardware Trojan signals should have large CC or CO values to avoid their detection as opposed to genuine signals that should have small CC or CO values to be testable, the 3-means algorithm effectively separates all genuine signals in one cluster and isolates all hardware Trojan signals in different clusters with considerable intercluster distances from the cluster of genuine signals. Considering the small CC or CO values for genuine signals and the relatively small size of hardware Trojans, the cluster of genuine signals is easily distinguishable as its centroid is the closest to the corner $\langle CC, CO\rangle = \langle 0, 0\rangle$ compared with the centroids of hardware Trojan clusters, and it contains the larger portion of circuit signals. Therefore, the COTD technique can effectively determine whether a circuit is hardware Trojan inserted without the need for any golden circuit.

5.2.3.1 Complexity: COTD vs. Some Existing Techniques

Usability of any hardware Trojan detection technique strongly depends on its time complexity. In this subsection, the time complexity of COTD is compared with HaTCh, VeriTrustX, and FANCIX techniques [2]. Both VeriTrustX and FANCIX techniques require monitoring of all sequential stages of design. VeriTrustX monitors the activation history of each entry in the truth table instead of terms in SOP/POS form of the Boolean function. Assuming m the number of circuit inputs, the time complexity of VeriTrustX is $O(2^m)$, that is, an exponential computation [2]. To compute the control value of each signal in a circuit, FANCIX needs to go through the each entry of developed truth tables and for all primary input combinations. This leads to the time complexity $O(m2^m)$ [2]. Assuming n represents the total number of wires in the circuit and d the worst case trigger signal dimension, HaTCh has the time complexity of $O((2n^2)^d)$ [2]. n and d may have a linear relationship and a logarithmic relationship with m, respectively. On the other hand, COTD consists of two steps in sequence: Step-1 determining controllability and observability values, and Step-2 executing k-mean clustering. Step-1 has the time complexity $O(n)$, and Step-2 has the time complexity of $O(nkdi)$ using the Lloyd's algorithm where n is the number of wires, d is the number of data dimensions, k the number of clusters which is set to 3 for COTD, and i the number of iterations needed until convergence. On data that do have a clustering structure, the number of iterations until convergence is often small, and results only improve slightly after the first dozen iterations. Therefore, the Lloyd's algorithm is often considered to be of "linear" time complexity in practice ($O(n)$). In total, the time complexity of COTD is $O(n) + O(n) = O(n)$.

Figure 5.8 shows a comparison of time complexity of different countermeasures. HaTCh-2in and HaTCh-4in show the complexities for the m-input AND gate circuit implementation with 2-input and 4-input AND gates, respectively. While HaTCh is sub-exponential in m and not exponential in m as for FANCIX and VeriTrustX, COTD outperforms and presents a linear relationship with m. The inconsiderable time complexity of COTD makes it a perfect choice for hardware Trojan detection in a gate-level netlist.

The COTD technique can be easily integrated into current integrated design flow as it is based on widely used commercial tools. Furthermore, the COTD technique does not require any pattern application; therefore, it outperforms the current existing techniques in terms of performance. As the COTD technique is

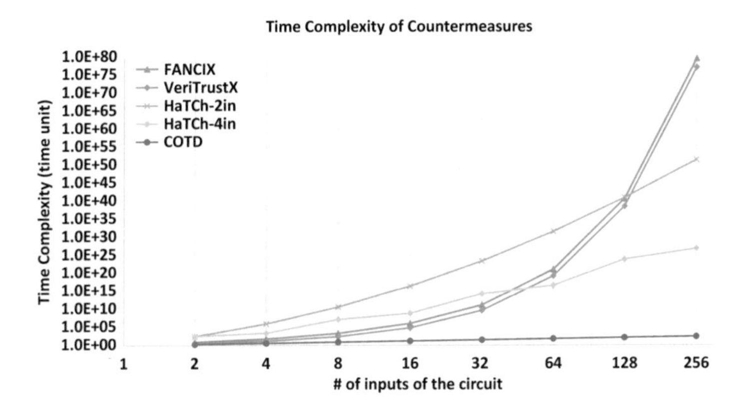

Fig. 5.8 Comparison of computational complexities of different hardware Trojan countermeasures [5]

based on controllability and observability analyses, any hardware Trojan circuit not connected to the main circuit would have infinite CC or CO values. As a result, they are separately clustered with high intercluster distance and can be identified easily. In addition, using the unsupervised clustering technique eliminates the need for a golden circuit as Hardware Trojan signals are clustered as signals with considerably low testability and with considerable intercluster distance from genuine signals.

5.2.3.2 Simulation Analyses

The COTD technique is applied to gate-level Trojan-inserted netlist provided on Trust-HUB [8] and some other circuits infected by Trojans proposed by DeTrust [11], always-on Trojans and HaTCh [2]. In two consecutive steps, COTD firstly obtains $CC0(s)$, $CC1(s)$, and $CO(s)$ for every signal s in a circuit using Synopsys TetraMAX [7], and then it executes unsupervised k-mean clustering with $k = 3$ to identify Trojan signals. Synopsys TetraMAX can report the SCOAP controllability and observability numbers in a set of numbers "CC0-CC1-CO." When each of these values exceeds the 254 limit that the program can track, it reports "*" for that value. In this work, "*" is replaced with 254 to execute COTD.

Table 5.2 presents the results of applying COTD to gate-level Trust-HUB benchmarks. While the first column indicates the name of benchmark, the second and third columns present the size of trigger and payload of inserted Trojan in terms of the number of gates. The next set of two columns presents the intercluster distances between the genuine-signal cluster and two Trojan-signals clusters. The next set of two columns indicates the amount of time required to obtain controllability and observability values and to perform clustering and identifying Trojan signals. The last two columns show the number of signals identified as genuine signals (No Genuine Signals) and the number of signals identified as Trojan signals (No

Table 5.2 Detecting hardware Trojans in Trust-HUB gate-level netlist using COTD [5]

Trojan	Trigger size	Payload size	Intercluster distance		Processing time (min:s)		No genuine signals (FN)	No Trojan signals (FP)
			Gen Clst-Tj. Clst 1	Gen Clst-Tj. Clst 2	Ctrl&Obsr	Clustering		
s38417-T100	11	1	316.97	255.69	00:03.47	00:0.08	5629(0)	12(0)
s38417-T200	11	4	246.66	247.41	00:02.89	00:0.06	5629(0)	15(0)
s38417-T300	15	29	431.11	246.87	00:02.66	00:0.07	5629(0)	44(0)
Ethernet-T700	12	1	298.14	247.86	00:11.97	00:1.04	102852(0)	13(0)
Ethernet-T710	12	1	283.50	247.45	00:09.96	00:1.10	102852(0)	13(0)
Ethernet-T720	12	1	298.14	247.80	00:08.98	00:1.10	102852(0)	13(0)
Ethernet-T730	12	1	298.14	247.75	00:09.75	00:1.08	102852(0)	13(0)
RS232-T1000	10	3	335.77	251.23	00:00.53	00:0.01	241(0)	13(0)
RS232-T1100	11	1	345.72	251.20	00:00.69	00:0.00	241(0)	12(0)
RS232-T1200	12	1	375.98	251.21	00:00.96	00:0.02	241(0)	13(0)
RS232-T1300	6	2	294.32	251.23	00:01.24	00:0.00	241(0)	8(0)
RS232-T1400	12	1	345.53	251.15	00:00.81	00:0.01	241(0)	13(0)
RS232-T1500	12	3	346.33	251.23	00:00.91	00:0.02	241(0)	15(0)
RS232-T1600	9	2	362.77	251.23	00:00.96	00:0.00	241(0)	11(0)
s15850-T100	26	1	331.02	250.80	00:00.83	00:0.03	2405(0)	27(0))
s35932-T100	13	2	435.86	250.52	00:00.80	00:0.07	6386(0)	15(0)
s35932-T300	12	24	250.51	251.03	00:00.95	00:0.06	6386(0)	36(0)
s38584-T200	126	1	431.49	252.14	00:0.97	00:0.08	7272(0)	127(0)
s38584-T300	1142	1	434.46	249.01	00:01.05	00:0.09	7272(0)	1143(0)
vga_lcd-T100	4	1	248.12	246.60	00:07.08	00:0.71	69832(0)	5(0)
wb_conmax-T100	11	4	184.42	153.41	00:02.02	00:0.24	22182(0)	15(0)

Trojan Signals). Further, false negative (FN) rate, the number of signals wrongly identified as genuine signals, and false positive (FP) rate, the number of signals wrongly identified as Trojan signals, are reported.

While the circuits have different sizes, and they contain Trojans with different sizes and functionality, all Trojans are successfully detected. For all benchmarks, signals are grouped in three clusters. All genuine signals are isolated in one cluster (Gen Clst), and all Trojan signals are divided into two clusters: one with high CC values (Tj. Clst1) and the other with high CO values (Tj. Clst2). The results in Table 5.2 highlight there is considerable intercluster distance between genuine signals and Trojan signals such that they can be easily distinguished. The average intercluster distances of Gen Clst and Tj. Clst1 and Gen Clst and Tj. Clst2 are about 328 and 245, respectively.

The processing timing analysis shows time to determine a circuit containing a Trojan has taken less than 14 s in the worst case. This result supports that the timing complexity of COTD is a linear order of the number of signals in the circuits as discussed in Sect. 5.2.3.1. Table 5.2 also presents the number of genuine and Trojan signals for each benchmark, and it shows both FN and FP rates are 0 for all cases. The zero rates indicate that there is no signal that is wrongly identified as a genuine signal while it is truly a Trojan signal and vice versa.

The results signify that realizing a stealthy hardware Trojan getting rarely activated would result in Trojans whose signals have significantly high controllability and observability values, that is, very difficult to control and observe. Otherwise, they are easily detectable. Interestingly, two gate-level Trojans in Trust-HUB benchmarks present low controllability and observability values, and they are frequently activated by only applying few hundred test patterns, shown in Table 5.3.

Shown in Table 5.3, s35932-T200 benchmark contains a Trojan whose trigger has 11 gates and its payload one gate. The analysis indicates that the maximum CC and CO for the genuine circuit and the Trojan circuit are $\langle 8.48, 12\rangle$ and $\langle 17.02, 20\rangle$, respectively. Comparing with Trojans in Table 5.2, the Trojan in s35932-T200 benchmark has very low controllability and observability values; therefore, it is observed the Trojan is activated 42 times by only applying random test patterns for 4261 test clock cycles. s38584-T100 benchmark also contains a small Trojan with low controllability and observability values. For this case, the maximum CC and CO values for the Trojan circuit are even less than those for the genuine circuit. Table 5.3 indicates that this Trojan is activated 21 times within only 3286 test clock cycles. Results in Table 5.3 emphasize that the controllability and observability values of Trojan signals cannot be small and close to those of genuine signals; otherwise Trojans frequently interfere the normal operation of a circuit.

The controllability and observability values of Trojan signals should be significantly high so that the Trojan remains hidden during circuit authentication. Table 5.4 presents the maximum and minimum of $|\langle CC, CO\rangle|$ for signals in the genuine circuit and Trojan circuit for Trust-HUB gate-level netlist, respectively. The results clearly indicate even the minimum magnitude of $|\langle CC, CO\rangle|$ for Trojan signals is much higher than the maximum magnitude of $|\langle CC, CO\rangle|$ for genuine signals. The larger this difference is, the stealthier a Trojan behaves. Almost all Trojans have a

Table 5.3 Gate-level Trojans in Trust-HUB with low controllability and observability values [5]

Trojan	Trigger size	Payload size	Genuine circuit		Trojan circuit		# of Trojan activation	# Test clock cycles
			Max CC	Max CO	Max CC	Max CO		
s35932-T200	11	1	8.48	12	17.02	20	42	4261
s38584-T100	8	1	35.35	32	7.07	9	21	3286

Table 5.4 Analysis of controllability and observability values for genuine and Trojan signals [5]

Trojan	Genuine circuit Max $\lvert\langle CC, CO\rangle\rvert$	Trojan circuit Min $\lvert\langle CC, CO\rangle\rvert$
s38417-T100	101.57	254.03
s38417-T200		254.05
s38417-T300		255.13
Ethernet-T700	88.01	254.01
Ethernet-T710		254.01
Ethernet-T720		254.03
Ethernet-T730		254.00
RS232-T1000	13.19	254.00
RS232-T1100		254.03
RS232-T1200		254.00
RS232-T1300		254.00
RS232-T1400		254.00
RS232-T1500		254.00
RS232-T1600		254.00
s15850-T100	86.15	254.00
s35932-T100	12.08	254.03
s35932-T300		254.00
s38584-T200	35.41	256.26
s38584-T300		254.13
vga_lcd-T100	42.05	256.00
wb_conmax-T100	98.34	142.39

minimum $\lvert\langle CC, CO\rangle\rvert$ value above 254 while the maximum $\lvert\langle CC, CO\rangle\rvert$ value for genuine signals ranges between 12 and 101.

Figure 5.9 depicts the results of unsupervised k-mean clustering with $k = 3$ for selected gate-level Trojans in Table 5.4. The figure shows genuine signals are localized in the bottom left corner of graph while Trojan signals are mainly located in three other corners of graph where CC, CO, or both are high. Furthermore, Fig. 5.9 depicts the significant intercluster distance between Trojan signals and genuine signals.

Among Trojans in Table 5.4, the maximum $\lvert\langle CC, CO\rangle\rvert$ value for genuine signals in wb_conmax-T100 benchmark is relatively close to the minimum $\lvert\langle CC, CO\rangle\rvert$ value for the Trojan inserted in this benchmark. While this Trojan is correctly detected, the low value of minimum $\lvert\langle CC, CO\rangle\rvert$ value for the Trojan compared to the other Trojans may expose its existence during the authentication phase. Random test patterns have been applied for 3779 test clock cycles to wb_conmax-T100 benchmark, and the Trojan becomes activated 5 times. While the Trojan in wb_conmax-T100 benchmark is stealthier than Trojans in s35932-T200 and s38584-T100 benchmarks analyzed in Table 5.3, having Trojan signals with low $\lvert\langle CC, CO\rangle\rvert$ values reveals the Trojan existence in a short time. This fact signifies that the minimum $\lvert\langle CC, CO\rangle\rvert$ values for Trojan signals should be considerably high

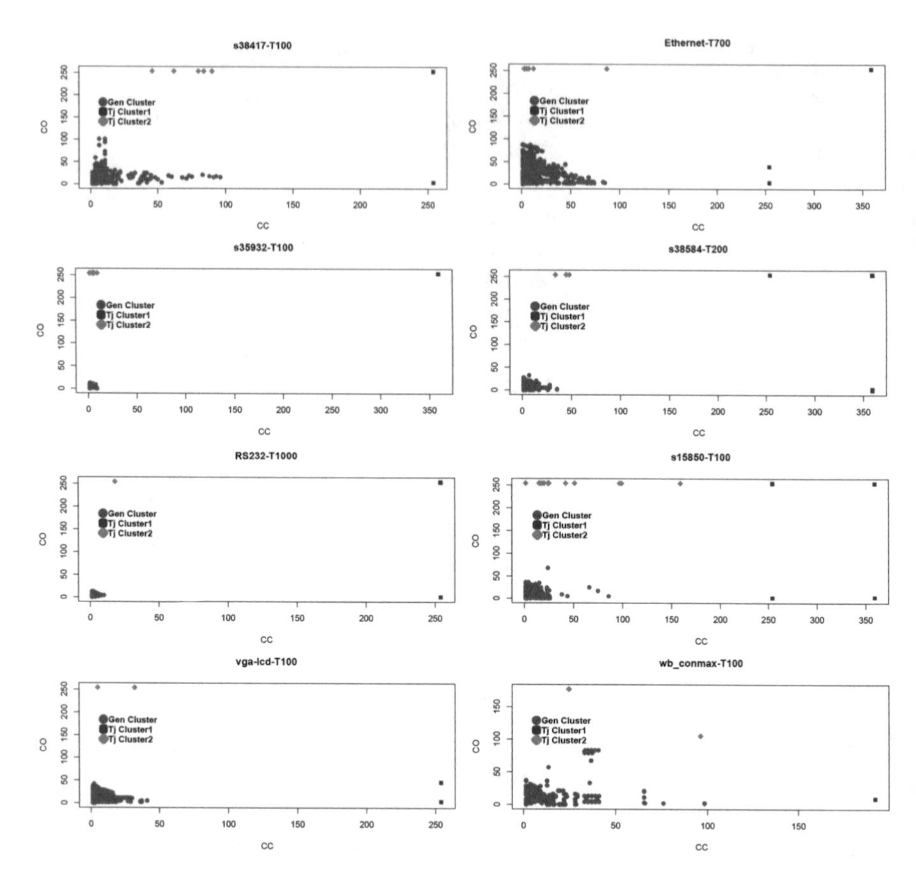

Fig. 5.9 Unsupervised k-mean clustering with $k = 3$ for selected gate-level Trojans on Trust-HUB [5]

to avoid Trojan activation during authentication time. On the other hand, COTD can catch Trojan signals as they stand in a cluster with considerably high intercluster distance from genuine signals.

Hardware Trojan Recovery It is so valuable to fully recover an inserted Trojan in a circuit to understand the purpose of adversary. Identifying Trojan trigger and Trojan payload circuitry presents detailed information about Trojan implementation. Further, it would also make it possible to determine (1) which signals are being used as inputs for the Trojan trigger, and (2) which signals are targeted by the Trojan payload.

After isolating Trojan signals using COTD, it is possible to fully recover a hardware Trojan inserted in a gate-level netlist. Figure 5.10 presents the proposed hardware Trojan recovery flow in a gate-level netlist. Trojan signals identified by COTD and Trojan-inserted circuit are used as inputs. The *Trojan Gates Identification* step extracts Trojan gates, and their input and output pins. With this

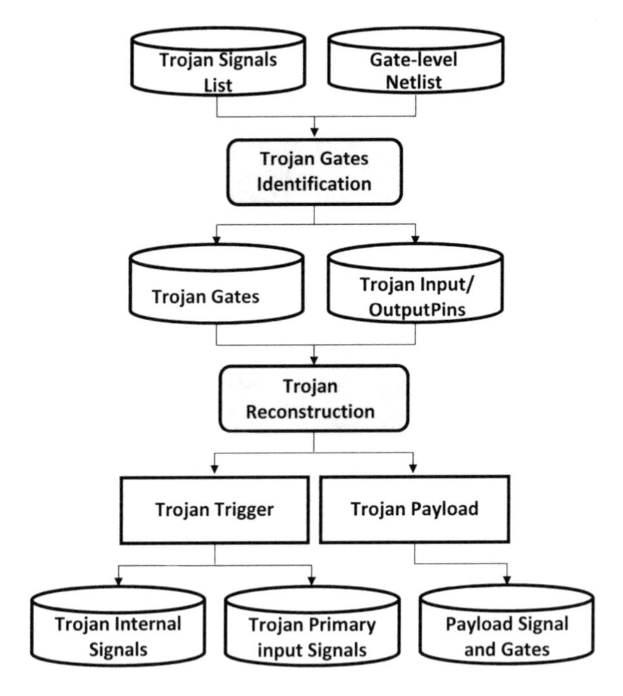

Fig. 5.10 The hardware Trojan recovery flow for gate-level netlist [5]

information, it is possible to execute the *Trojan Reconstruction* step. In this step, Trojan trigger and Trojan payload circuitry are restored. Signals connected to Trojan gates' pins are obtained, and the interconnection between Trojan gates is reconstructed. Further, it is determined which signals from the main circuit are being used as Trojan triggering signals, and which signals in the main circuit are attacked by the Trojan payload. Any signal that drives a Trojan gate and is not driven by any Trojan gate is identified as a Trojan triggering signal. Any signal that is not a Trojan signal but passing through a Trojan gate is identified as a Trojan payload signal. Any gate whose one of inputs is a payload signal composes the Trojan payload circuitry. The remaining Trojan gates compose the Trojan trigger circuitry. The hardware Trojan recovery flow is implemented in Synopsys' design compiler and only consists of about 100 lines, and its complexity is an order of the number of Trojan signals. The flow is being applied to all gate-level netlist on Trust-HUB, and all Trojans are successfully recovered.

As a sample, Fig. 5.11 presents the output of flow for the s38417-T100 Trojan on Trust-HUB. While Fig. 5.11 shows a part of report, the report includes (1) Trojan gates for each of which input and output pins with their connected nets are reported, (2) Trojan internal signals are distinguished, (3) Trojan triggering signals are detected, and (4) Trojan payload gates, along with targeted nets are identified. Comparing the report in Fig. 5.11 with the Trojan inserted in s38417-T100 circuit shows the proposed hardware Trojan recovery flow can perfectly extract the inserted Trojan. While a Trojan circuit can be completely isolated, it is also possible to clean up the Trojan-inserted circuit.

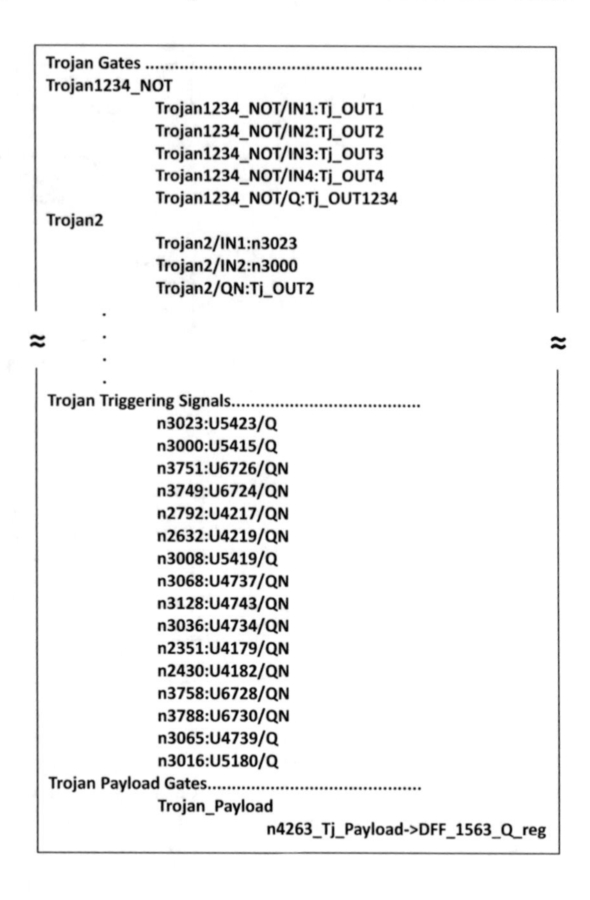

Fig. 5.11 The restored s38417-T100 Trojan [5]

5.3 Conclusions

The extensive practice of integrated circuit horizontal design flow and demand for untrusted third-party firm intellectual properties underline the need for hardware Trojan detection at the gate level. This chapters reviewed some of major existing techniques and discussed their effectiveness. Majority of techniques rely on switch activity analyses while such analyses do not seem applicable to large and complex designs such as system of chips. Eliminating false positive and negative rates in a reasonable time is a major existing challenge for all techniques. While some techniques seem promising, there is need for considerable evaluations on industrial-size circuits to determine their effectiveness.

References

1. B. Çakir, S. Malik, Hardware Trojan detection for gate-level ICs using signal correlation based clustering, in *Proceedings of the 2015 Design, Automation and Test in Europe Conference and Exhibition (DATE)* (2015), pp. 471–476
2. S.K. Haider, C. Jin, M. Ahmad, D.M. Shila, O. Khan, M.V. Dijk, HaTCh: a formal framework of hardware Trojan design and detection, Cryptology ePrint Archive, Report 2014/943 (2014)
3. W. Hu, D. Mu, J. Oberg, B. Mao, M. Tiwari, T. Sherwood, R. Kastner, Gate-level information flow tracking for security lattices. ACM Trans. Des. Autom. Electron. Syst. **20**(1), 2:1–2:25 (2014)
4. M. Oya, Y. Shi, M. Yanagisawa, N. Togawa, A score-based classification method for identifying hardware-Trojans at gate-level netlists, in *Proceedings of the 2015 Design, Automation and Test in Europe Conference and Exhibition (DATE)* (2015), pp. 465–470
5. H. Salmani, COTD: reference-free hardware trojan detection and recovery based on controllability and observability in gate-level netlist. IEEE Trans. Inf. Forensics Secur. **12**(2), 338–350 (2017)
6. G.A.F. Seber, *Multivariate Observations* (Wiley, Hoboken, 1984)
7. Synopsys's TetraMAX. http://www.synopsys.com/Tools/Pages/default.aspx
8. Trust-HUB benchmarks. https://www.trust-hub.org/taxonomy
9. A. Waksman, M. Suozzo, S. Sethumadhavan, FANCI: identification of stealthy malicious logic using Boolean functional analysis, in *Proceedings of the 2013 ACM SIGSAC Conference on Computer and Communications Security (CCS)* (2013), pp. 697–708
10. C. Wu, L. Wang, X. Wen, *VLSI Test Principles and Architectures: Design for Testability*. The Morgan Kaufmann Series in Systems on Silicon (Morgan Kaufmann Publishers, San Francisco, 2006)
11. J. Zhang, F. Yuan, Q. Xu, DeTrust: defeating hardware trust verification with stealthy implicitly-triggered hardware Trojans,in *Proceedings of the 2014 ACM SIGSAC Conference on Computer and Communications Security (CCS)* (2014), pp. 153–166

Chapter 6
Circuit Vulnerabilities to Hardware Trojan at the Layout Level

6.1 Circuits at the Layout Level

The high complexity of modern designs, the shortened time-to-market window, and the cost restriction of final product highly drive the horizontal design process. The third-party intellectual properties (3PIPs) are widely used in the forms of soft, firm, and hard IPs to expedite the design process. As making and maintaining of a fabrication site are very expensive, external fabs are preferred to reduce the product cost. While a cooperative design supply chain is widely practiced in the interest of economy, the appearance of different external parties across the design development provides opportunities to malicious ones to tamper with the design by inserting extra circuits called hardware Trojans.

A circuit layout is the result of the physical design synthesis. The physical design synthesis obtains a synthesized netlist together with a technology library and delivers a valid placement layout. The physical design synthesis flow contains four major steps to meet design functional and parametric specifications: placement, buffering and resizing, clock-tree synthesis, and routing. Placement is an essential step of the flow and assigns the exact location for various circuit components within the chip's core area. Placement is twisted with cell resizing and buffering that are essential for timing and signal integrity satisfaction. Clock-tree synthesis and routing follow completing the physical design process. When all design requirements are met, the circuit layout is signed off for fabrication.

When a circuit layout becomes available to an untrusted foundry for fabrication, it can insert hardware Trojans that leave minimum footprints. To reduce a Trojan activation probability, its triggers are connected to nets with low transition probabilities. However, to minimize Trojan impacts, other parameters including availability of whitespace, unused routing channels for Trojan placement should be studied. Furthermore, the existence of nets with low transition probability on noncritical paths should be investigated. Therefore, it is necessary to develop a systematic approach to analyze the susceptibility of circuit layout to hardware Trojan insertion.

H. Salmani, *Trusted Digital Circuits*, https://doi.org/10.1007/978-3-319-79081-7_6

6.2 Motivation

Like genuine cells of a circuit, hardware Trojan cells need to be placed and routed to realize Trojan functionality. To minimize a hardware Trojan's footprint, empty regions (whitespaces) in a circuit layout with available routing channels in metal layers above the regions are suitable candidates for Trojan cells placement.

For example, ITC99's b15 benchmark [1] consists of 3296 cells, and Fig. 6.1 shows the distribution of whitespace as a unit of INVX0 (the smallest gate in SAED_EDK90nm library) across the b15 layout after synthesizing the benchmark using the Synopsys's SAED_EDK90nm library at 90 nm technology node [2]. The layout is divided into tiles with the area A_T equal to W^2 where W is the width of the largest cell in the design library, that is, $A_T = 560.7424\,\mu\text{m}^2$.

The detailed results indicate that the densest area of b15 layout has 84.83 units of INVX0 around the center of the layout, and the average density is 53.71 units of INVX0. With the average size of 41.41 units of INVX0, whitespaces are more prevalent in areas closer to the layout boundaries, and plenty of areas with low density are also available across the b15 layout.

Figure 6.2 presents normalized average available routing channel for 9-metal layer implementation of b15 benchmark. With the maximum and average routing congestion of 0.36 and 0.15, respectively, a significant amount of unused routing channel is available to be used for routing Trojan cells without requiring to modify the original circuit routing. The results indicate the availability of 0.84 unused routing channel on average per unit of A_T. Regions of circuit layout containing whitespace and unused routing channel are highly susceptible to Trojan insertion.

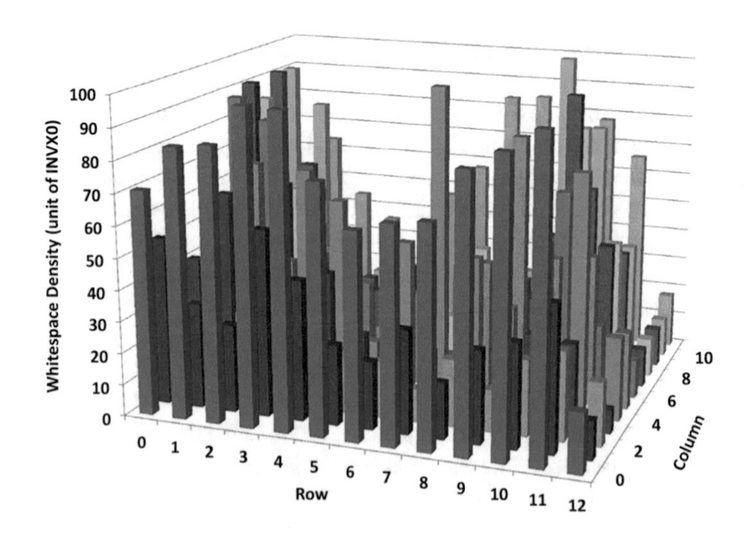

Fig. 6.1 The distribution of whitespace across the b15 layout

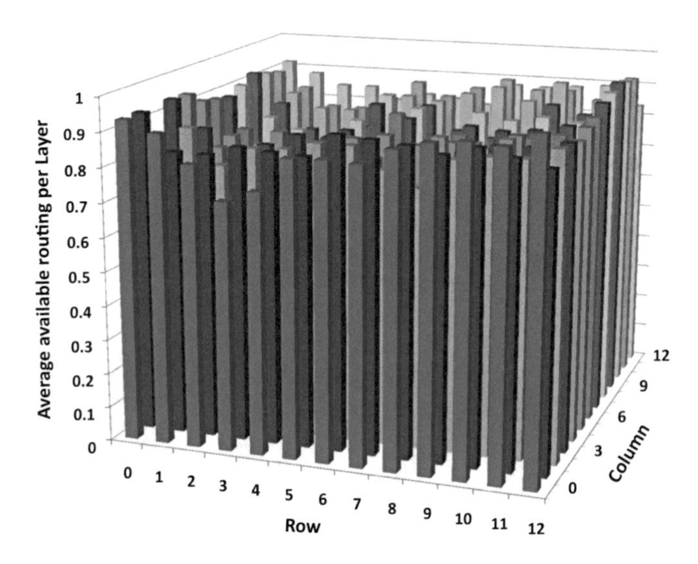

Fig. 6.2 Average available routing area at different metal layers of b15 layout

While availability of whitespaces and routing channels reduces efforts in inserting a hardware Trojan, such existence does not necessarily promise a stealthy implementation of hardware Trojan circuit. For example, if one or some of triggering signals of a hardware Trojan are connected to cells located in considerable distance, Trojan induced delay (TID) due to wiring connection may reveal Trojan existence. Therefore, more detailed analysis is required to consider Trojan impact on design specifications.

To reside away from delay-based detection techniques, Trojan triggers and payloads may be connected to nets on noncritical paths whose delay is less than 75% of the critical path of circuit. Figure 6.3 shows the distribution of noncritical paths across the b15 layout. The analysis shows that there are about 17 nets on noncritical paths per region, on average. Furthermore, there is tendency toward the center of circuit such that the figure increases to about 24 nets, on average. The existence of a considerable number of nets on noncritical paths per region brings vulnerability to Trojans resistant to delay-based detection techniques.

Power-based Trojan detection techniques examine circuit power consumption to distinguish Trojan contribution. To minimize Trojan activity, Trojan triggers might be connected to nets with low transition probabilities. Figure 6.4 shows the distribution of nets with transition probability smaller than a transition probability threshold (P_{th}) equal to 1E−04 in b15 benchmark. The figure shows the considerable number of nets with low transition probabilities in some regions. For example, the region 160 in row 12 and column 4 contains 12 nets with transition probabilities smaller than 1E−04. This detailed analysis shows, in total, there are only 358 nets (11% of total nets in b15 benchmark) with transition probability smaller than 1E−04, and these nets can be flagged as possible Trojan triggers.

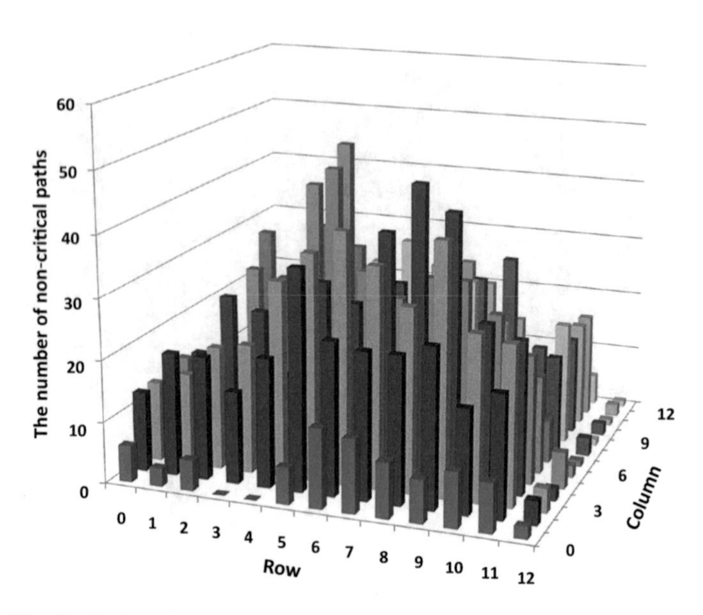

Fig. 6.3 The distribution of noncritical path in b15 benchmark

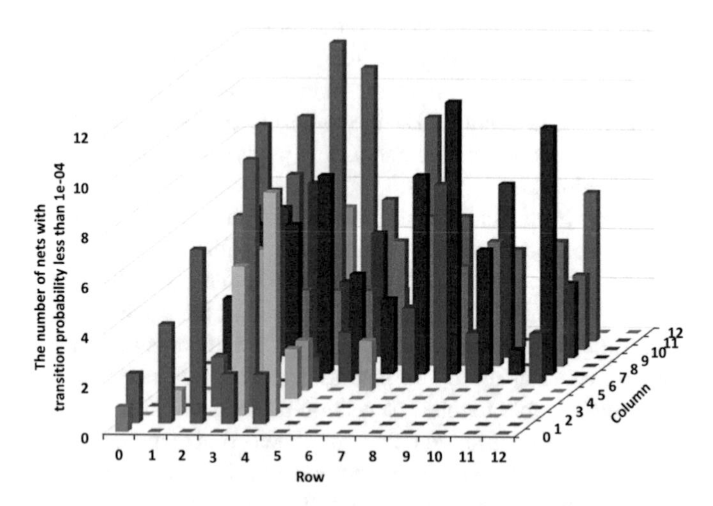

Fig. 6.4 The distribution of nets with transition probabilities smaller than $P_{th} = 1e{-}04$ in b15 benchmark

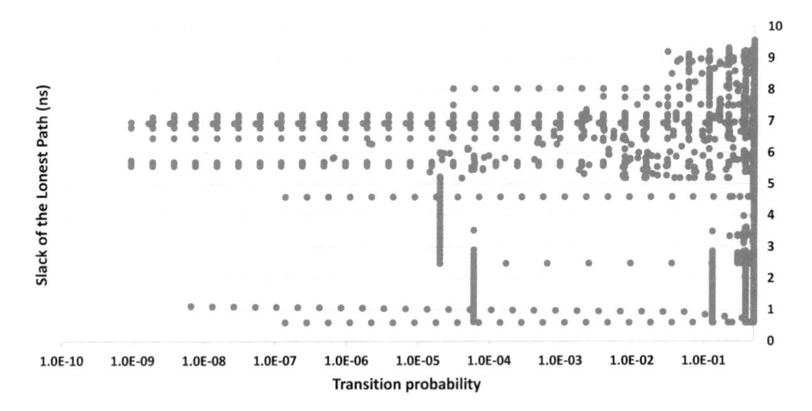

Fig. 6.5 The slack time of the longest path passing through a net versus the net transition probability for all nets of b15 benchmark

Some Trojan detection techniques perform multiple-parameter analyses to capture Trojan effects on several parameters, like power and delay, at the same time. Ideally to escape delay and power-based detection techniques, Trojan Triggers and Payloads should have two characteristics: (1) To be connected to nets with low transition probabilities (2) To be placed on paths among which the path with maximum delay is yet a noncritical path These two conditions ensure the minimum contribution of a Trojan into circuit power consumption and the least impact on circuit performance characteristic. Figure 6.5 shows nets' transition probability versus the slack time of the longest path passing through each net for b15 benchmark. The figure clearly shows there are a significant number of nets with low transition probability on noncritical paths. These nets can be used to provide Trojan trigger and payload with minimum impact on design characteristics. Figure 6.6 presents the distribution of nets whose transition probability is smaller than 1E−04 and the delay of the longest path passing through is less than 75% of the critical path. The figure reveals there are regions with a considerable number of nets that can be used for Trojan trigger and payload. For example, in the region 111 at row 8 and column 7, there are 11 nets that can realize a Trojan trigger with the activation probability of 6.96E−58 while Trojan induced delay would be inconsiderable as input triggers are on noncritical paths.

Analyzing b15 benchmark signifies that the vulnerability analysis of circuit layout to hardware Trojan insertion is required to perform a detailed analysis and isolate regions that are more susceptible to hardware Trojan insertion. Furthermore, quantifying the vulnerability of each region of circuit layout to different types of Trojans may reduce design efforts to prevent hardware Trojan insertion in backend design stages. And, it may provide Trojan detection guidance based on the vulnerability of a region.

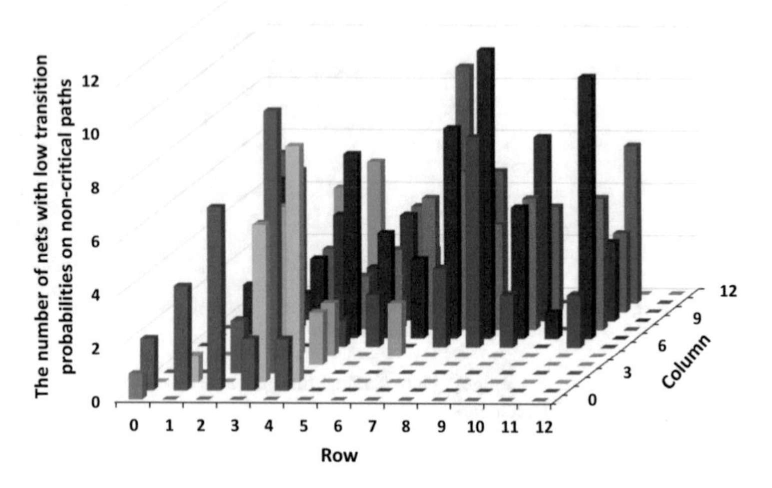

Fig. 6.6 The distribution of nets with low transition probabilities on noncritical paths for b15 benchmark

6.3 Layout Vulnerability Analysis Flow

A physical design tool takes a synthesized netlist and associated technology library information and performs placement and routing considering design constraints such as performance, size, and manufacturability. Cells are typically placed in rows and their interconnections are realized through metal layers above the cells. Large circuits such as system-on-chips take larger areas for placement and require more metal layers for routing to meet circuit constraints, besides circuit functionality. However, a final circuit layout may contain a considerable amount of whitespaces in substrate and empty routing channels in metal layers above the substrate. These empty spaces can be used by an untrusted foundry to place and route hardware Trojan cells with minimum impact on circuit specification. Figure 6.7 shows a flow to analyze the vulnerability of a circuit layout to hardware Trojan insertion.

6.3.1 Cell and Routing Analyses

A gate-level synthesized netlist, along with design constraints and technology library information, is fed into a physical design tool for placement and routing. The circuit layout, the output of physical design, shows gates' location and their detail wiring through metal layers. In addition to the circuit layout, an updated design netlist with circuit parasitic parameters is obtained. The proposed flow includes two novel steps: the *Cell Analysis* and the *Routing Analysis* to study cell distribution and

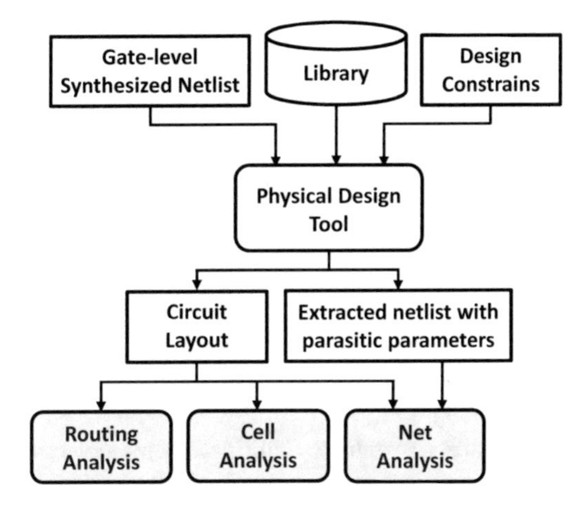

Fig. 6.7 The layout vulnerability analysis flow

routing congestion. The Cell Analysis step screens the circuit silicon to extract cells' location. It then determines whitespaces distribution and their size. The Routing Analysis step extracts used routing channels in each metal layer and determines unused ones.

After obtaining the circuit layout, the Cell Analysis obtains the circuit size and collects placed cells and their coordination by screening the circuit layout. Using these information, whitespaces in the circuit substrate are identified. Any whitespace whose area size is greater than that of the smallest cell in the technology library is considered a potential location for one or more Trojan cells insertion. The Cell Analysis also obtains the distribution of cells and whitespaces across the layout. The Routing Analysis collects used and unused routing channels in metal layers above the substrate. Available routing channels can potentially be used for Trojan cells interconnection and their connections to the main circuit. Similar to the Cell Analysis, the Routing Analysis also collects the distribution of used and empty routing channels in all metal layers. After determining whitespace and unused routing channels distributions of a circuit layout, the vulnerability of a region of circuit layout to hardware Trojan cells placement is defined as

$$V(r) = WS(r) \times UR(r) \tag{6.1}$$

where $V(r)$ indicates the vulnerability of region r, $WS(r)$ the normalized whitespace of region r, and $UR(r)$ the normalized unused routing channels of region r. It is expected that Trojan cells are inserted in regions with high $V(r)$ where there are equally high $WS(r)$ and $UR(r)$.

While being inserted in a region with high V may not guarantee Trojan detection avoidance, to remain hidden from delay-based detection techniques, Trojans should be inserted in regions with enough whitespace and empty routing channels and tapped to nets on noncritical paths. The value of vulnerability to delay-resistance Trojans ($V_{Td}(r)$) in the region r can be defined as

$$V_{Td}(r) = V(r) \times N_{NC}(r) \tag{6.2}$$

where $N_{NC}(r)$ is the number of noncritical path in the region r and $V(r)$ is the vulnerability of region r as defined by Eq. (6.1).

To stand the power-based detection technique, Trojans should be connected to nets with low transition probabilities and placed in regions with enough whitespace and unused routing channel. The value of vulnerability to power-resistance Trojans ($V_{Tp}(r)$) in the region r can be defined as

$$V_{Tp}(r) = V(r) \times N_{LP}(r) \tag{6.3}$$

where $N_{LP}(r)$ is the number of nets with transition probability smaller than a predefined P_{th} in the region r, and $V(r)$ is the vulnerability of region r as defined by Eq. (6.1).

Trojans resistant to multiparameter detection techniques should be placed in regions with empty space and unused routing channels and connected to nets with low transition probabilities located on noncritical paths. The value of vulnerability to power and delay-resistance Trojans ($V_{Tdp}(r)$) in region r can be defined as

$$V_{Tdp}(r) = V(r) \times N_{NC\&LP}(r) \tag{6.4}$$

where $N_{NC\&LP}(r)$ is the number of nets with transition probability less than a predefined P_{th} on noncritical paths in the region r and $V(r)$ is the vulnerability of region r as defined by Eq. (6.1).

6.3.2 Net Analysis

In general, a functional Trojan consists of two parts: a Trojan trigger and a Trojan payload. The Trojan trigger is connected to some internal nets, and the Trojan payload restitches some other nets. The Trojan trigger determines conditions under which the Trojan payload propagates erroneous values into the main circuit. The Net Analysis performs a comprehensive analysis of each net in a circuit. The analysis determines the transition probability of each net in the circuit. With incorporating the circuit layout information, it is possible to obtain the distribution of transition probability across the circuit layout. Using a timing analysis tool, the slack distribution of worst paths passing through a net can be obtained. Nets with low transition probability located on noncritical paths are suitable candidates for Trojan trigger inputs.

A hardware Trojan may deem to leak secret information from a design or cause denial of service or functionality change. To launch such attacks, a Trojan payload may target signals with low testability to reduce Trojan detection probability. Testability is a relative measure of the effort or cost of testing a logic circuit, and it can be used to identify nets with poor testability [3]. From the security perspective, nets with low testability less frequently determine the values of circuit primary outputs and its states; therefore, their manipulation highly remains hidden. Testability analysis can be performed by calculating the controllability and observability of each signal. Controllability reflects the difficulty of setting a signal line to a required logic value and observability reflects the difficulty of propagating the logic value of the signal line to observation points. The Sandia Controllability/Observability Analysis Program (SCOAP) is the most popular testability program that measures testability of each signal s in a circuit logic based on several numerical values including $CC0(s)$—combinational 0-controllability of s, $CC1(s)$—combinational 1-controllability of s, and $CO(s)$—combinational observability of s. These combinational testability measures roughly determine the number of signals that must be manipulated in order to control or observe s from primary inputs or at primary outputs. The values of controllability measures range between 1 and ∞, while the values of observability measures range between 0 and ∞. As a boundary condition, the $CC0$ and $CC1$ values of a primary input are set to 1, and the CO value of a primary output is set to 0. From the security perspective, signals with low controllability and low observability are more susceptible to be targeted by a Trojan payload. The susceptibility of a signal s to Trojan payload attack ($Payload(s)$) based on their testability measures is defined as

$$Payload(s) = \frac{1}{\sqrt{CC0^2(s) + CC1^2(s)} \times CO(s)}. \tag{6.5}$$

A signal s with high $Payload(s)$ is a signal which is highly observable or highly controllable. Otherwise, a signal s with low $Payload(s)$ is difficult both to control and observe; therefore, such signals are highly susceptible to Trojan Payload attack.

In conclusion, the layout-level vulnerability analysis flow identifies regions of a circuit that are more vulnerable to Trojans resistant to delay-based, power-based, and multiparameter-based detection techniques. Furthermore, the vulnerability of a region is quantified, and this provides a detailed and fair comparison between different circuit implementations. With such knowledge, it is possible to incorporate effective prevention techniques with the least impact on main design specifications. In addition, the flow may provide insightful guidance for authenticating circuits after manufacturing.

6.4 Simulation Results

The layout vulnerability analysis flow is applied on several benchmarks including b15, b19, and Ethernet benchmarks [1, 4]. The Synopsys's SAED_EDK90nm library at 90 nm technology node [2] is used, and 9 metal layers are considered for routing. The Routing Analysis, Cell Analysis, and Net Analysis are developed using the tool control language (Tcl), and they can be easily integrated into any commercial electronic design automation (EDA) tool. In this paper, the Routing Analysis and Cell Analysis are performed using Synopsys's IC Compiler and the Net Analysis using Synopsys's Design Compiler, Synopsys's PrimeTime, and Synopsys's Tetramax.

To determine N_{NC} for each region, the circuit layout is obtained and corresponding circuit netlist and circuit parasitics after physical design using Synopsys's IC Compiler. Circuit netlist and circuit parasitics are passed to Synopsys's PrimeTime. Meantime, the location of each net in the circuit is obtained in Synopsys's IC Compiler and passed to Synopsys's PrimeTime. In Synopsys's PrimeTime, the delay of the longest path passing through each net is obtained. If this delay is less than 75% of the circuit clock period, the net is flagged noncritical. By knowing that a net is critical or noncritical and combining with the location of net, it is possible to determine N_{NC} in each region of circuit layout. To determine N_{LP}, the transition probability of every net in the circuit is determined in Synopsys's Design Compiler using a program in TcL that is already made available by author on www.trust-hub.org website. In a similar manner, net transition probability and net location are combined to determine N_{LP} in each region. Finally, $N_{NC\&LP}$ value is obtained by combing net location, net transition probability, and the delay of the longest path passing through a net. To evaluate the significance of the proposed metrics, several Trojans are implemented and the metrics are calculated and compared against Trojans' impact of design characteristics.

The significance of the flow is that it is developed based on existing commercial tools; therefore, it benefits from their high-performance execution. The complexity of determining N_{NC}, N_{LP}, and $N_{NC\&LP}$ is $O(n)$ where n is the number of nets in a circuit. The complexity of determining the transition probability of every net in the circuit is effectively $O(n)$, as well. Collecting the whitespace and unused routing channels is performed in Synopsys's IC Compiler, and the procedure is a function of the size of circuit layout. Therefore, the proposed vulnerability flow has a linear relation with circuit size and can be applied on large industrial circuits.

In the following, the vulnerability analysis for b15 benchmark is explained in detail, and the same analysis for the other circuits is performed and the results are only presented.

Case Study 1: b15 For example, the average available white space and unused routing channel are about 41 unit of INVX0 and 0.84 per layer, respectively. After performing layout vulnerability analysis flow, Fig. 6.8 presents the vulnerability of b15 benchmark to Trojan insertion at the layout level as defined by Eq. (6.1). The average vulnerability is about 0.46 and about 40% of regions have V above 0.5. These signify considerably high susceptibility of the layout to Trojan insertion.

Figure 6.9 shows the vulnerability of b15 benchmark to Trojans resilient to delay-based detection techniques across the layout. The results indicate the region 95 in Row 7 and Column 4 is the most susceptible region to delay-resistance Trojan with $V_{Td}(95) = 20.44$, where $N_{NC}(95) = 28$ and $V(95) = 0.73$. Interestingly, the adjacent region 94 in Row 7 and Column 3 has considerably higher number of noncritical paths ($N_{NC}(94) = 40$); however, the region 94 has low whitespaces or unused routing channels, $V(94) = 0.21$. Therefore, the susceptibility of region 94 is much lower with $V_{Td}(94) = 8.4$.

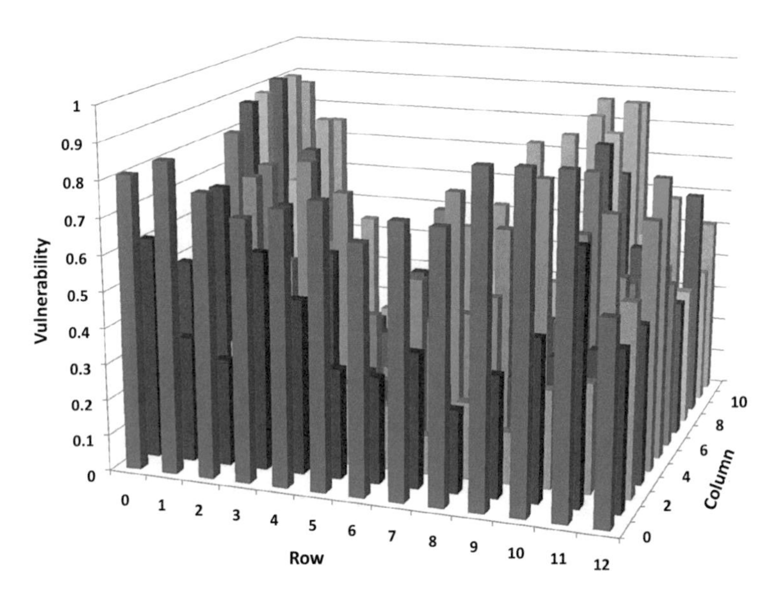

Fig. 6.8 The vulnerability of b15 benchmark to Trojan insertion at the layout level ($V(r)$)

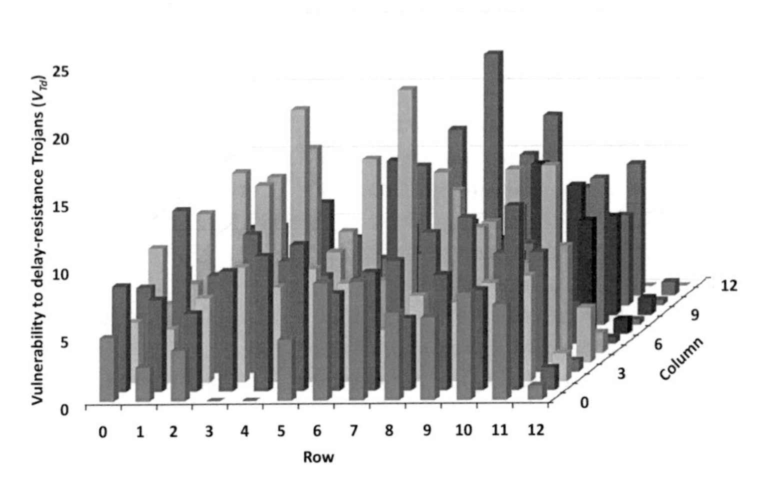

Fig. 6.9 Vulnerability of b15 benchmark to delay-based Trojans

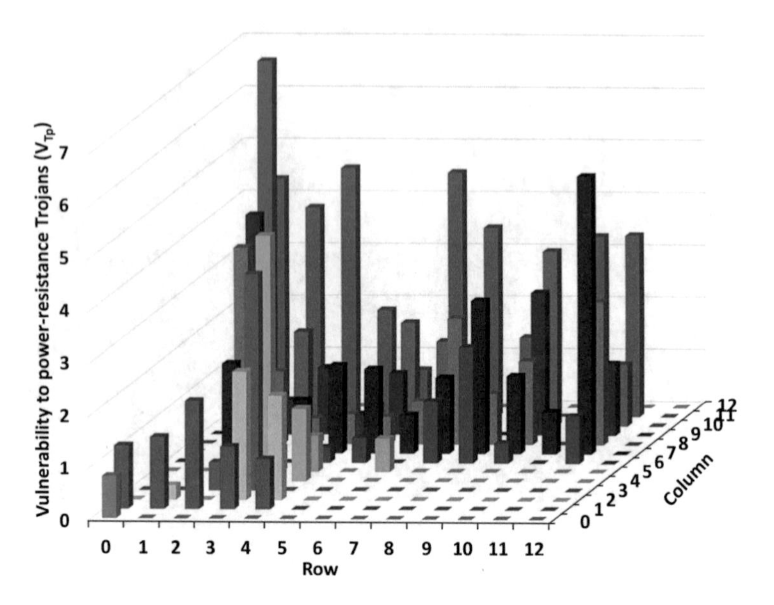

Fig. 6.10 Vulnerability of b15 benchmark to power-based Trojans with $P_{th} = 1e{-}04$

The vulnerability of b15 benchmark to Trojans resilient to power-based detection techniques across the layout is shown in Fig. 6.10 with $P_{th} = 1e{-}04$. The results show that the region 147 in row 11 and column 4 has the maximum vulnerability to power-resistance Trojan with $V_{Tp}(147) = 7.71$ where $V(147) = 0.70$ and $N_{LP}(147) = 11$.

Comparing results for delay-resistance Trojans and power-resistance Trojans reveals that b15 benchmark is more susceptible to delay-resistance Trojans as the maximum V_{Td} is greater than the maximum V_{Tp}. Furthermore, regions with the maximum V_{Td} and V_{Tp} are different such that the most susceptible region to delay-resistance Trojan is region 95 and the most susceptible region to power-resistance Trojan is region 147 for b15 benchmark.

Figure 6.11 shows the vulnerability of b15 benchmark to Trojans resilient to multiparameter power and delay Trojan detection techniques. The results reveal that the region 150 in row 11 and column 7 is the most susceptible region to Trojans resilient to power and delay-based detection techniques with $V_{Tdp}(150) = 5.32$ where $V(150) = 0.53$ and $N_{NC\&LP}(150) = 10$. The analysis signifies even with using multiparameter Trojan detection techniques, it is possible to implement a Trojan whose full activation probability can be about 1−E40 while there is still considerable whitespace and unused routing channels in the region 150. This analysis flags regions that are highly susceptible to Trojan insertion; therefore, it can effectively limit Trojan investigation into a limited number of regions. The detailed analysis for b15 benchmark shows that 21 regions out of 169 regions have V_{Tdp} above 3 that indicates the moderate existence of regions with considerable

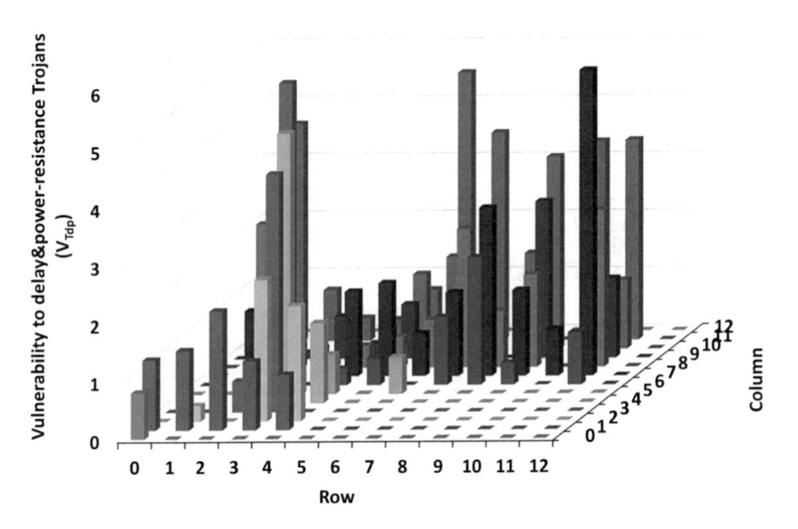

Fig. 6.11 Vulnerability of b15 benchmark to Trojans resilient to multiparameter Trojan detection techniques

whitespace and unused routing channels and a considerable number of nets with low transition probability on noncritical paths.

The detailed results for b15 benchmark show that the percentage of regions with V_{Td} above 5, 10, and 15 are 75%, 17%, and 3%, respectively. The percentage of regions with V_{Tp} above 2, 4, and 5 are 14%, 4%, and 0.6%, and the percentage of regions with V_{Tdp} above 2, 4, and 5 are 12%, 2%, and 0%. The results emphasize that the b15 benchmark has a higher percentage of regions vulnerable to Trojans resilient to delay-based Trojan detection techniques. Further, by using multiparameter power and delay Trojan detection techniques these percentages significantly drop.

To reduce Trojan detection probability, Trojan payload may target nets with low testability. Table 6.1 lists 10 nets with the lowest $Payload$ value in b15 benchmark. For the signal n3689, $\langle CC0, CC1, CO\rangle$ values are $\langle 34, 97, 133\rangle$ and $Payload(n3689)$ is the lowest (7.31E−05). The testability analysis indicates the signal n3689 has low controllability for both 0 and 1 and low observability. Table 6.1 also indicates in which region of a circuit layout a signal is located (r) and what its $V(r)$ is. Although the signal n3689 has a very low $Payload$ value, it is located in the region 69 ($r = 69$) with a low vulnerability ($V(69) = 0.17$); therefore, the signal n3689 may not be a good candidate for an attack by Trojan Payload. The next signal with the lowest Payload value is the signal add_0_root_add_371_3/n17 with $Payload = 8.13\text{E}{-}05$. Although this signal is located in a region with higher vulnerability ($V(36) = 0.25$), it is located on a critical path whose slack time is 0.60 ns. Therefore, this signal may not be a good candidate for Payload attack as

Table 6.1 The net analysis for Trojan payload in the b15 layout

Net	CC0	CC1	CO	Payload	Slack (ns)	r	V
n3689	34	97	133	7.31E−05	5.80	69	0.17
add_0_root_add_371_3/n17	2	251	49	8.13E−05	0.60	36	0.25
add_0_root_add_360_3/n15	2	245	50	8.16E−05	4.59	23	0.77
add_0_root_add_371_3/n16	2	229	49	8.91E−05	0.60	35	0.11
n3682	1	66	169	8.96E−05	5.38	69	0.17
add_0_root_add_360_3/n14	2	223	50	8.97E−05	4.59	35	0.11
add_0_root_add_371_3/n15	2	208	49	9.81E−05	0.60	35	0.11
add_0_root_add_360_3/n13	2	202	50	9.90E−05	4.59	34	0.28
N3424	34	34	196	1.06E−04	5.80	69	0.17
add_0_root_add_371_3/n14	2	188	49	1.09E−04	0.60	34	0.28
add_0_root_add_360_3/n6	2	182	50	1.10E−04	4.59	34	0.28

Table 6.2 The vulnerability analysis for Ethernet benchmark

Vuln.		Power-resistant Tj. Vuln.				Delay-resistant Tj. Vuln.				Power and Delay-resistant Tj. Vuln.			
r	V	r	N_{LP}	V	V_{Tp}	r	V	N_{NC}	V_{Td}	r	V	$N_{NC\&LP}$	V_{Tdp}
0	0.889	499	15	0.690	10.350	514	0.750	33	24.734	499	0.690	11	7.590
182	0.872	498	19	0.418	7.938	579	0.575	39	22.438	496	0.422	8	3.378
770	0.854	434	25	0.287	7.182	499	0.690	29	20.010	498	0.418	8	3.342
2752	0.845	435	13	0.520	6.765	665	0.735	26	19.120	432	0.371	6	2.225
235	0.841	497	16	0.365	5.845	435	0.520	36	18.734	435	0.520	4	2.082
527	0.840	579	9	0.575	5.178	684	0.643	28	17.993	53	0.563	3	1.689
2880	0.836	433	14	0.323	4.519	385	0.664	27	17.919	665	0.735	2	1.471
171	0.835	665	5	0.735	3.677	496	0.422	42	17.735	497	0.365	4	1.461
546	0.830	495	5	0.724	3.619	449	0.645	25	16.113	495	0.724	2	1.448
3200	0.824	496	7	0.422	2.956	367	0.664	24	15.937	434	0.287	5	1.436

any modification of it high likely impacts design performance. Table 6.1, instead, indicates that the signal add_0_root_add_360_3/n15 with $Payload = 8.16\text{E}{-05}$ is located in the region $r = 23$ with high vulnerability ($V(r) = 0.77$), and it is located on noncritical paths (slack = 4.59 ns). Therefore, it can be concluded that the signal add_0_root_add_360_3/n15 in b15 benchmark is highly susceptible to Trojan Payload attack.

Case Study 2: Ethernet The vulnerability analysis flow is also applied to Ethernet benchmark, and the results are presented in Table 6.2. In four sets of columns, Table 6.2 shows the vulnerability (labeled Vuln.), the vulnerability to power-resistant Trojans (labeled Power-resistant Tj Vuln.), the vulnerability to delay-resistant Trojans (labeled Delay-resistant Tj Vuln.), and the vulnerability to power and delay-resistant Trojans (labeled Power and Delay-resistant Tj Vuln.).

The Vuln. column set presents regions of Ethernet layout with the highest vulnerability due to existence of considerable whitespace and unused routing channels. The next set of columns, Power-resistant Tj Vuln., shows the most vulnerable regions of Ethernet benchmark to Trojan resistant to power-based detection techniques where P_{th} is set to 1E−04. Interestingly, none of these regions are listed in Vuln. column set so that it signifies the importance of V_{Tp} metric. The third set of columns presents the vulnerability of Ethernet benchmark to Trojans resistant to delay-based detection techniques. While the clock period is set to 20 ns and the critical path has about 0.37 ns slack time, Ethernet benchmark presents significant vulnerability to Trojan resistant to delay-based detection techniques. Comparing V_{Tp} and V_{Td} values indicates that Ethernet is more vulnerable to Trojans resistant to delay-based detection techniques rather than power-based detection techniques. The comparison also indicates that the regions with the highest vulnerability to Trojans resistant to delay-based detection techniques and Trojans resistant to power-based detection techniques are different, and the magnitude of vulnerability to Trojans resistant to delay-based detection techniques is much higher. The last set of Columns, Power and Delay-resistant Tj. Vuln., presents the most vulnerable regions of Ethernet layout to Trojans resistant to power and delay-based detection techniques. While multiparameter detection techniques increase Trojan detection probability, V_{Tdp} values for the 10 most vulnerable regions also indicate small vulnerability of Ethernet benchmark to Trojans resistant to power and delay-based detection techniques.

Studying V_{Tp}, V_{Td}, and V_{Tdp} values for the 10 most vulnerable regions of Ethernet benchmark shows that the region 499 at row 7 and column 51 of Ethernet layout is significantly vulnerable to Trojans resistant to power-based, delay-based, and power and delay-based detection techniques. Interestingly the region 44 is not one of the 10 regions with highest whitespace and unused routing channels. This analysis provides insightful guidance for design authentication such that it prioritizes regions for circuit based on their vulnerability to different types of Trojans.

Table 6.3 presents 10 nets with the lowest $Payload$ value in Ethernet benchmark. The analysis shows the nets have low Payload value and they do not pass any critical

Table 6.3 The net analysis for Trojan payload in the Ethernet layout

Net	CC0	CC1	CO	Payload	Slack (ns)	r	V
n145918	44	8	44	5.08E−04	14.06	1631	0.61
n145968	44	8	43	5.20E−04	14.10	1124	0.50
n145942	36	9	45	5.99E−04	13.04	1057	0.52
n145943	32	6	49	6.27E−04	12.66	994	0.60
n21414	36	1	39	7.12E−04	18.63	496	0.42
n21392	1	41	34	7.17E−04	18.63	496	0.42
n145971	36	9	37	7.28E−04	14.26	1565	0.50
n145972	32	6	41	7.49E−04	13.87	1629	0.55
n146014	28	1	46	7.76E−04	16.32	432	0.37
n145917	15	6	73	8.48E−04	10.56	1630	0.49

Table 6.4 The vulnerability analysis for b19 benchmark

Vuln.		Power-resistant Tj. Vuln.				Delay-resistant Tj. Vuln.				Power and Delay-resistant Tj. Vuln.			
r	V	r	N_{LP}	V	V_{Tp}	r	V	N_{NC}	V_{Td}	r	V	$N_{NC\&LP}$	V_{Tdp}
25	0.960	2513	20	0.697	13.940	2466	0.782	39	30.493	2513	0.697	20	13.940
2715	0.953	185	17	0.643	10.932	1740	0.810	34	27.536	185	0.643	17	10.932
2706	0.951	1860	14	0.756	10.582	2301	0.793	34	26.971	1538	0.697	13	9.062
2708	0.951	1538	13	0.697	9.062	954	0.643	40	25.702	1868	0.493	16	7.891
2716	0.949	1868	16	0.493	7.891	1824	0.738	34	25.079	1382	0.693	11	7.628
50	0.949	1382	11	0.693	7.628	1823	0.648	38	24.615	1862	0.687	11	7.558
2709	0.944	1862	11	0.687	7.558	2316	0.825	28	23.106	1604	0.687	11	7.556
1325	0.943	1604	11	0.687	7.556	1840	0.786	29	22.792	939	0.741	10	7.409
2749	0.943	939	10	0.741	7.409	2455	0.653	34	22.203	336	0.643	11	7.077
2714	0.942	2000	10	0.735	7.351	1856	0.735	30	22.042	186	0.643	11	7.076

path. Furthermore, they are located in regions with relatively high V. Therefore, the listed nets are highly vulnerable to Trojan payload attack as their manipulation would leave inconsiderable impact on Ethernet circuit's characteristics.

Case Study 3: b19 Table 6.4 presents the 10 most vulnerable regions in b19 benchmark after applying the vulnerability analysis flow. The Vuln. column set indicates significant amount of whitespace and unused channel in b19 layout that can be potentially used for Trojan insertion. The Power-resistant Tj Vuln. column set presents the most vulnerable regions of b19 layout to Trojans resistant to power-based detection techniques. While none of these regions are recognized in the Vuln. column set, they show relatively high V_{Tp} due to considerably high N_{LP} and V. The Delay-resistant Tj Vuln. column set points out much higher vulnerability of b19 layout to Trojans resistance to delay-based detection techniques. While the clock period of b19 circuit is set to 11 ns, there is a considerable number of nets in each region that do not pass through any critical path; therefore, with large V and N_{NC}, b19 layout has large V_{Td}. The vulnerability of b19 circuit to Trojans resistant to delay and power-based detection techniques is presented in the Power and Delay-resistant Tj Vuln. column set. The analysis shows that the majority of regions with the highest V_{Tp} also have high V_{Tdp}, that is many nets in these regions are on noncritical paths.

Table 6.5 shows the vulnerability of b19 benchmark to Trojan payload attack. Interestingly, 10 nets with the lowest *Payload* in b19 benchmark are located on paths whose slack time is very close to 25% of b19's clock period, that is, 2.75 ns. If these nets are manipulated by Trojan payload attack, there is possibly of detection using faster-than-clock-speed testing techniques. Therefore, the analysis suggests an attacker should avoid these nets and find some other nets, although with higher *Payload* value, located on noncritical path.

As VTp, VTd, and VTdp metrics are based on circuit implementation, they make it possible to perform inter-circuit analysis and determine which implementation is

Table 6.5 The net analysis for Trojan payload in the b19 layout

Net	CC0	CC1	CO	Payload	Slack (ns)	r	V
P1_mult_1420/CARRYB_23__0_	45	248	234	1.70E−05	2.81	2264	0.44
P1_mult_1421/CARRYB_23__0_	45	248	234	1.70E−05	2.91	1647	0.64
P1_mult_1420/CARRYB_24__1_	47	220	251	1.77E−05	2.82	2263	0.43
P1_mult_1421/CARRYB_24__1_	47	220	251	1.77E−05	2.92	1594	0.69
P1_mult_1420/CARRYB_22__0_	43	238	223	1.85E−05	2.81	2264	0.44
P1_mult_1421/CARRYB_22__0_	43	238	223	1.85E−05	2.91	1647	0.64
P1_mult_1420/CARRYB_23__1_	45	212	243	1.90E−05	2.82	2263	0.43
P1_mult_1421/CARRYB_23__1_	45	212	243	1.90E−05	2.92	1646	0.49
P1_mult_1420/CARRYB_21__0_	41	228	215	2.01E−05	2.82	2264	0.44
P1_mult_1421/CARRYB_21__0_	41	228	215	2.01E−05	2.91	1647	0.64

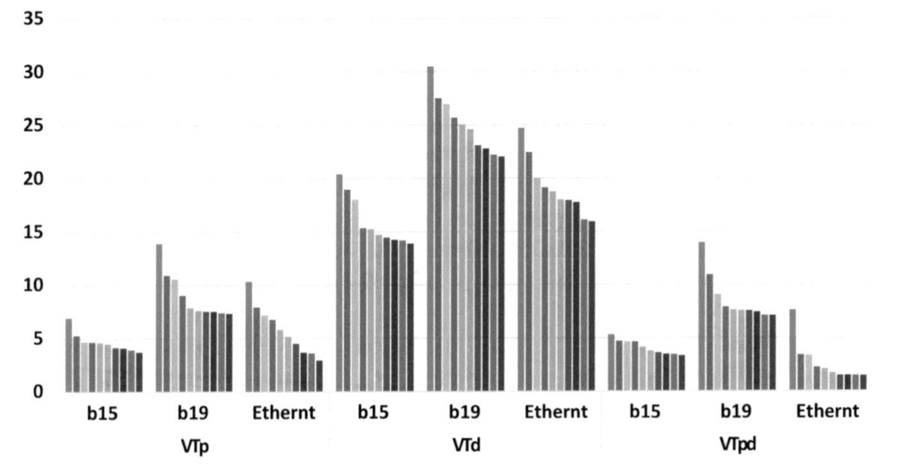

Fig. 6.12 Inter-circuit vulnerability analysis for the 10 most vulnerable regions in b15, b19, and Ethernet benchmarks

less or more vulnerable against different attacks. Figure 6.12 compares b15, b19, and Ethernet benchmarks against V_{Tp}, V_{Td}, and V_{Tdp} for the most 10 vulnerable regions. The results show that all three circuits are more vulnerable to Trojans that are resilient to delay-based Trojan detection techniques with average 15.95, 25.05, and 19.07 for b15, b19, and Ethernet benchmarks, respectively. And b19 is more vulnerable than others. Results of V_{Tp} indicate that benchmarks are less vulnerable to power-based Trojan detection techniques with 4.69, 8.99, and 5.80, on average, for b15, b19, and Ethernet benchmarks, respectively. The vulnerability of benchmarks is further reduced against multiparameter Trojan detection techniques to 4.11, 8.61, and 2.61, on average, for b15, b19, and Ethernet benchmarks, respectively. Results indicates while benchmarks are significantly vulnerable to Trojans resilient to delay-based detection techniques, their vulnerabilities are significantly reduced by 3.88X, 2.90X, and 7.30X using multiparameter Trojan detection, for b15, b19, and Ethernet benchmarks, respectively. These detailed analyses make it possible to fairly compare different design implementations against different attacks.

Trojan Analysis Based on the Trojan taxonomy [5, 6], a Trojan can have a loose distribution or a tight distribution. With a loose distribution, Trojan cells are placed relatively far from each other. On the other hand, Trojan cells are placed relatively close to each other in a limited area in a tight distribution. To study the impact of Trojan distribution on its power consumption, a Trojan comparator with 11 cells consisting of six NOR2X1 gates and five AND2X1 gates is implemented both with loose and tight distributions in b15 circuit. To realize the distributions, the tight Trojan is dispersed over 900 μm^2 and the loose Trojan over 25,110 μm^2. For both implementations, the same inputs are used and Table 6.6 shows the impact of Trojan distribution on its power consumption.

Table 6.6 shows Trojan induced capacitance on Trojan trigger inputs in a loose distribution is considerably higher than that in a tight distribution. The same holds true with Trojan internal capacitance as Trojan gates are located in a farther distance from each other. Table 6.6 indicates that the total Trojan induced delay is $0.134pF$ and $0.358pF$ in tight and loose distribution, respectively. Considering that both Trojans are connected to the same triggering inputs, Trojan power consumption is 0.2205 and 0.2236 μW for tight and loose distribution, respectively. The comparator with the loose distribution consumes about 1.4% more power as the total induced capacitance in the loose distribution is about 2.6X more than that in the tight distribution.

To study the V_{Tp} metric, a Trojan is inserted in three regions with different V_{Tp} values. The Trojan is a 3-bit synchronous counter whose enable signal is driven by a 3-input AND gate. When the output of the AND gate is '1', the counter is incremented by 1; otherwise, the previous value is kept. The three Trojans are labeled Counter_R1, Counter_R2, and Counter_R3, and Table 6.7 presents $V(r)$

Table 6.6 The impact of Trojan distribution on power consumption for 12-input comparator

Distribution	Tight	Loose
Dispersion (μm^2)	900	25110
Trojan inputs capacitance (pF)	0.092	0.205
Trojan internal capacitance (pF)	0.042	0.153
Capacitance of Trojan nets (inputs + internals) (pF)	0.134	0.358
Power (W)	2.205E−07	2.236E−07

Table 6.7 The V_{Tp} metric and Trojan power consumption

Trojan	Counter_R1	Counter_R2	Counter_R3
$V(r)$	0.7737	0.835	0.1835
$V_{Tp}(r)$	6.963	3.34	2.0189
# of partial activity	14	5	9
Average power consumption (W)	4.94E−007	5.29E−007	5.67E−007

and $V_{Tp}(r)$ for each Trojan. The $V(r)$ value for Counter_R2 is the highest with 0.835 that indicates the existence of considerable whitespace and unused routing channels. However, Table 6.7 shows that Counter_R1 has the highest $V_{Tp}(r)$, that is, Counter_R1 is in a region with higher susceptibility to power-resistance Trojans. The same test pattern set is applied to the three Trojans and '# of partial activity' in Table 6.7 presents the number of switching activity inside each Trojan circuit, and the last row of Table 6.7 presents the average power consumption of each Trojan. The results interestingly indicate although Counter_R3 has the lowest # of partial activity equal to 9, its average power consumption is the highest (0.567 μW). On the other hand, Counter_R1 with # of partial activity equal to 14 has the lowest average power consumption (0.494 μW). The measurements signify the importance of V_{Tp} metric. The larger value of $V_{Tp}(r)$ indicates that a region r is more susceptible to power-resistance Trojans as they consume less power and would be more challenging to detect them. In comparison with Counter_R3, Counter_R1 experiences more switching activities and consumes less power compared with Counter_R3 and its $V_{Tp} = 6.963$. On the other hand, the Counter_R3 Trojan consumes more power so that it would be easier to detect it, that is less susceptible to power-resistance Trojans with smaller $V_{Tp}(= 2.0189)$.

To hide Trojan impact on design delay characteristics, Trojan should be connected to nets on noncritical paths. Furthermore, to reduce Trojan induced delay (TID), the distance between cells driving Trojan triggers and the cells of Trojan trigger should be as small as possible. Otherwise, the long wiring would considerably impact the delay of Trojan triggering signals. To study the impact of distance, one Trojan gate with the input capacitance of about 1.194X of SAED_EDK90 INVX0 is inserted in three different locations (Tj_Loc1, Tj_Loc2, and Tj_Loc3 in Table 6.8) and connected to the same input. Table 6.8 shows the distance of the Trojan gate from its driving gate. The Tj_Loc1 Trojan is about 7.12 μm far from its driving gate, and this causes TID of about 0.008 ns. The second Trojan, Tj_Loc2, is placed in a distance about 33.50 μm, and it increases the delay of driving signal by about 0.024 ns. To realize a Trojan connection, an attacker may route the connection among different layers to reduce the wiring length; however, a long wiring can still expose Trojan if it is connected to a critical path. The Tj_Loc3 Trojan is placed in about 265.32 μm far from the driving cell. Considering that the width of the largest cell in Synopsys's SAED_EDK90nm library is about 23 μm, the Tj_Loc3 Trojan is placed as far as about 11 the largest cell in the library. Due to this considerable distance, a significant increase of TID of about 0.180 ns is observed. The results in Table 6.8 signify that the distance of a Trojan from its driving cells is deterministic, and a long connection may result in considerable TID.

Table 6.8 The impact of distance of a Trojan trigger input from its driving gate on Trojan induced delay (TID)

Trojan	TID (ns)	Distance (μm)
Tj_Loc1	0.008	7.12
Tj_Loc2	0.024	33.50
Tj_Loc3	0.180	265.32

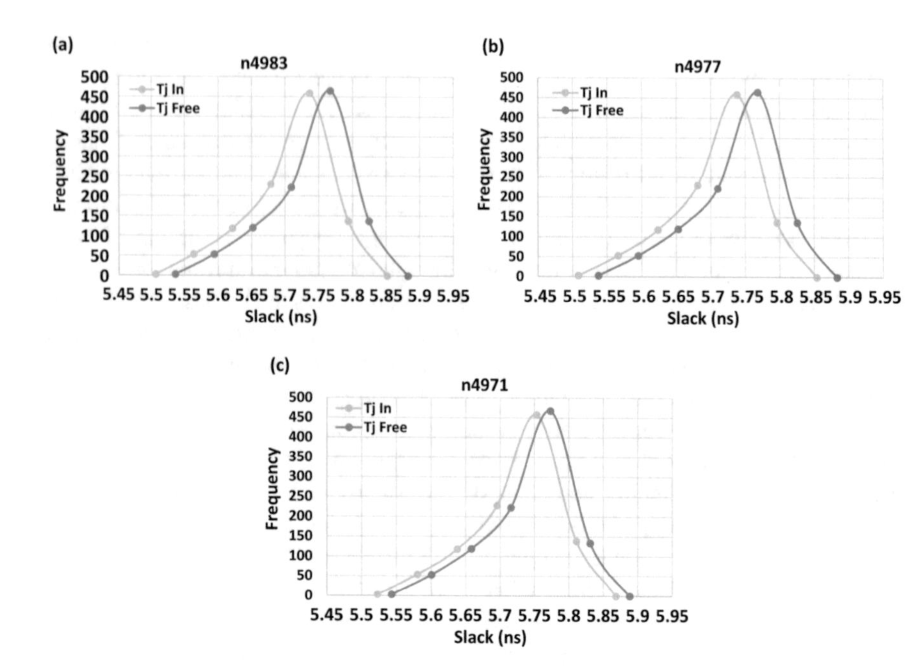

Fig. 6.13 The slack distribution of 1000 worst paths passing through each triggering signal of a 3-bit synchronous counter Trojan inserted in Region 33 of b15 benchmark with $V_{Td}(33) = 2.71$. (**a**) Slack distribution for $n4983$ (**b**) Slack distribution for $n4977$ (**c**) Slack distribution for $n4971$

The V_{Td} metric determines vulnerability of a region of circuit to Trojans resistant to delay-based detection techniques. A 3-bit synchronous counter Trojan is inserted in the region 33 at Column 7 and row 2 with $V_{Td}(33) = 2.71$ ($V(33) = 0.1430$ and $N_{NC}(33) = 19$). The enable signal of counter is provided by 3-bit AND gates driven by the triggering signals $n4983$, $n4977$, and $n4971$ of b15 benchmark. Figure 6.13 shows the slack distribution of the 1000 worst paths passing through each triggering signal before and after Trojan insertion. While the clock period of b15 benchmark is set to 15 ns, the average slack for the $n4983$, $n4977$, and $n4971$ signals respectively are 5.678, 5.680, and 5.695 ns with the same standard deviation of 0.057 ns in the Trojan free circuit-the paths are long paths but not critical paths as their slack times are greater than 25% of the clock period. After inserting the Trojan, reductions of about 0.030, 0.028, and 0.020 ns are observed in the slack for the 1000 worst paths passing through the $n4983$, $n4977$, and $n4971$ signals, respectively. The reduction is indeed TID that causes the increase in the paths' delay. Although the value of $V_{Td}(33)$ is small, the 3-bit synchronous counter Trojan is also small and its impact on circuit delay characteristics may not be caught using conventional delay testing methods. The 3-bit synchronous counter Trojan taps to only three triggering signals and to reduce its impact on delay characteristics, the 3-bit AND gate needs to be placed close to the triggering signals.

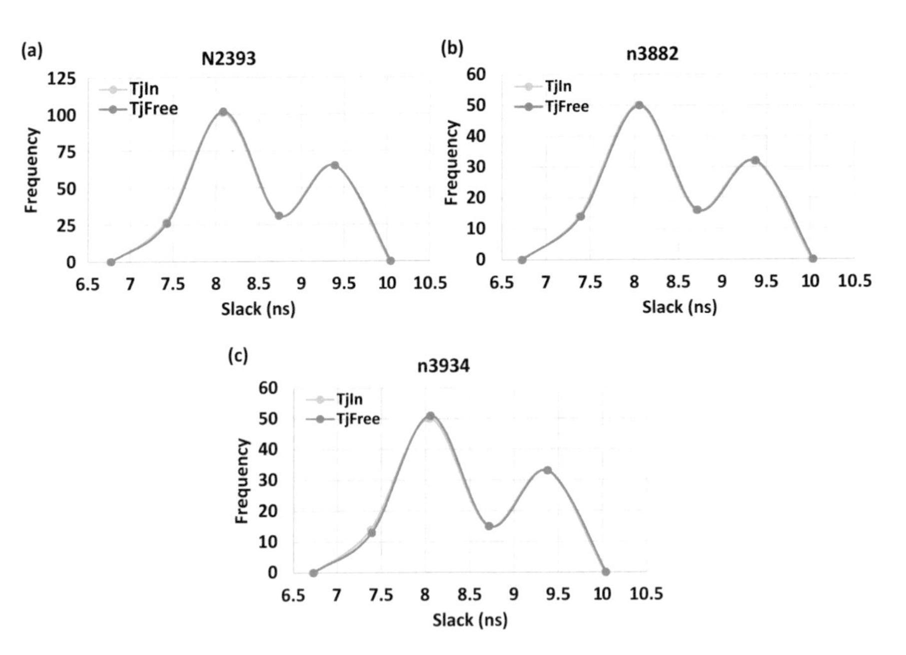

Fig. 6.14 The slack distribution of all paths passing through each triggering signal of a 3-bit synchronous counter Trojan inserted in Region 95 of b15 benchmark with $V_{Td}(95) = 20.42$. (**a**) Slack distribution for $N2393$ (**b**) Slack distribution for $n3882$ (**c**) Slack distribution for $n3934$

The same 3-bit synchronous counter Trojan is placed in the region 95 at Column 4 and Row 7 with $V_{Td}(95) = 20.42$ ($V(95) = 0.7293$ and $N_{NC}(95) = 28$). The $V_{Td}(95)$ value indicates the considerably high vulnerability of region 95 to Trojans resistant to delay-based detection techniques. Figure 6.14 shows the slack distribution of all paths on which the triggering signals of the counter pass through-the $N2393$, $n3882$, and $n3934$ signals. As the number of paths passing through each signal is small, their slack distribution does not show a normal distribution. The analysis shows the minimum and maximum slacks are about 6.5 and 10 ns for the three triggering signals in Trojan-free circuit. The analysis also shows the TID for the three signals is about 0.01 ns after Trojan insertion, and Fig. 6.14 indicates it is almost impossible to distinguish Trojan-inserted and Trojan-free circuits based on the circuit delay characteristics. Comparing to the TID of the counter Trojan in the region 33, the smaller TID of the Trojan in the region 95 is attributed to the higher $V_{Td}(95)$ compared with the smaller $V_{Td}(33)$ so that the Trojan can be placed and routed with much less impact on original design characteristics. Therefore, the V_{Td} metric can effectively represent the vulnerability of a region to Trojans resistant to delay-based detection techniques.

The V_{Tdp} metric determines the vulnerability of a region to Trojans which are resistant to both delay-based detection techniques and power-based detection techniques. One 3-bit synchronous counter Trojan and one 12-bit comparator separately inserted into b15 benchmark and Figs. 6.15 and 6.16 respectively show their delay

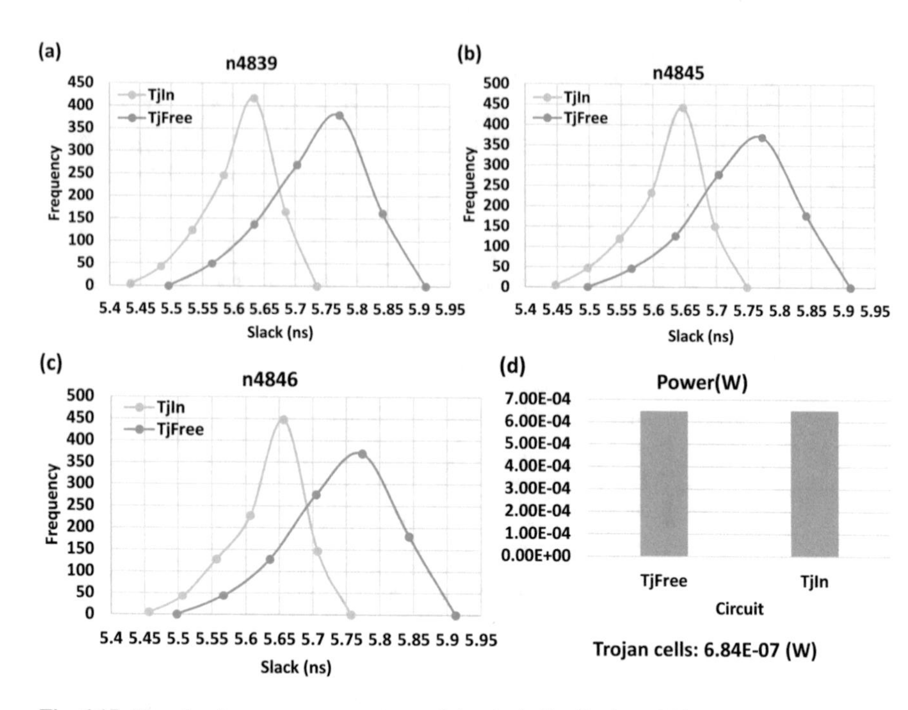

Fig. 6.15 The circuit power consumption and the slack distribution of 1000 worst paths passing through each triggering signal of a 3-bit synchronous counter Trojan inserted in Region 42 of b15 benchmark with $V_{Tdp}(42) = 4.149$. (**a**) Slack distribution for $n4839$ (**b**) Slack distribution for $n4845$ (**c**) Slack distribution for $n4846$ (**d**) Trojan cells' power

and power impacts. The counter Trojan is inserted in the region 42 at Column 3 and Row 3 with $V_{Tdp}(42) = 4.149$ ($V(42) = 0.4149$ and $N_{NC\&LP}(42) = 10$). Figure 6.15a–c show the slack distribution of the 1000 worst paths passing through the three triggering signals of the counter. The analysis indicates the amount of TID for the three signals is about 1 ns, on average, and the minimum slack for each signal after Trojan insertion still is so large that the worst path does not become a critical path. Figure 6.15d also presents the very small power consumption of the Trojan circuit ($\approx$ 6.84E−07 W), and the circuit power consumption before and after Trojan insertion has almost remained the same, about 6.49E−04 W. Therefore, the 3-bit synchronous counter Trojan may remain hidden from both delay- and power-based Trojan detection techniques.

A similar analysis is performed for 12-bit comparator inserted in the neighboring region 43 at Column 4 and Row 3 with $V_{Tdp}(43) = 4.7035$ ($V(43) = 0.78$ and $N_{NC\&LP}(43) = 6$). The slack distributions of three selected inputs of the comparator with the minimum slacks are presented in Fig. 6.16a–c. The amount of TID is 0.018 ns, on average, and the delay of the worst path is not large enough to be considered a critical path. In Fig. 6.16d, the Trojan power consumption is very small ($\approx$ 4.44E−07 W). With a small impact on circuit delay characteristics and power consumption, the comparator Trojan may also remain hidden. Comparing the 3-bit

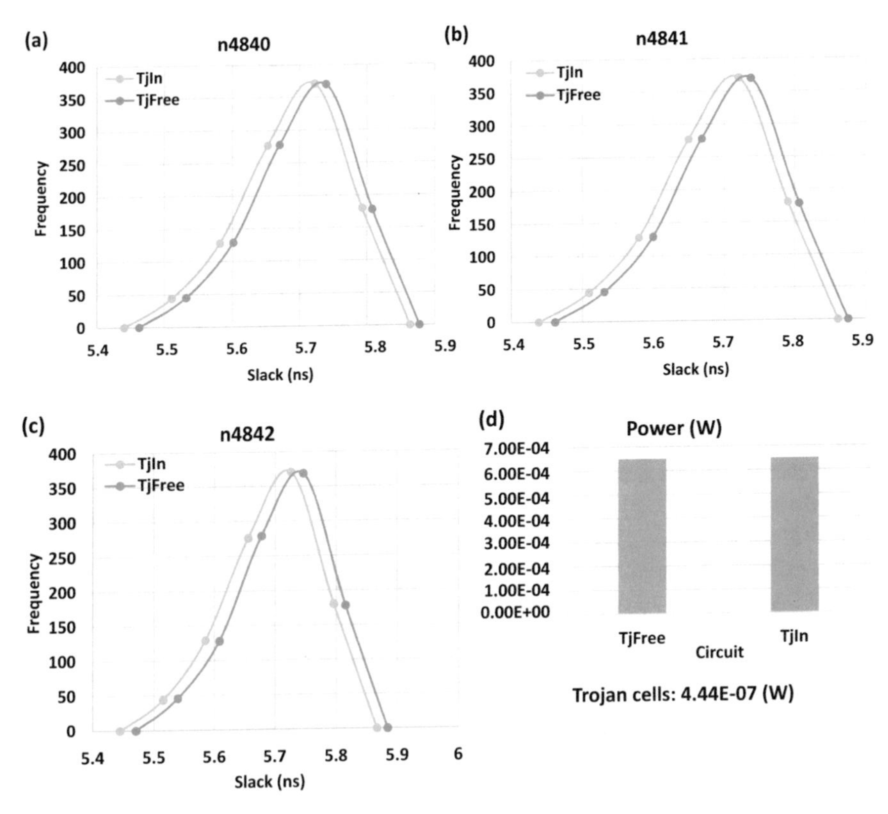

Fig. 6.16 The circuit power consumption and the slack distribution of 1000 worst paths passing through three selected triggering signal of a 12-bit comparator Trojan with minimum slack inserted in Region 43 of b15 benchmark with $V_{Tdp}(43) = 4.7035$. (**a**) Slack distribution for $n4840$ (**b**) Slack distribution for $n4841$ (**c**) Slack distribution for $n4842$ (**d**) Trojan cells' power

synchronous counter and 12-bit combinational comparator indicates that the counter consumes more power than the comparator although the size of comparator circuit is larger. This is attributed to the fact the counter is a sequential circuit and the clock inputs of its flip-flops are connected to circuit clock. Furthermore, TID for the comparator circuit is smaller than that of the counter which is because of the higher V_{Tdp} value of the comparator. Therefore, the V_{Tdp} metric can effectively identify regions vulnerable to Trojans resilient to delay and power-based techniques.

6.5 Conclusions

The existence of untrusted manufacturers necessitates a novel layout vulnerability analysis flow to determine susceptibility of a circuit layout to hardware Trojan insertion. While there has not been that many considerable and comprehensive work toward vulnerability at the layout level, this chapter reviewed one existing work that proposed layout vulnerability analysis flow. The proposed layout vulnerability analysis flow enabled to determine the susceptibility of a circuit layout to hardware Trojan insertion. Based on a circuit layout's placement and routing information, metrics were introduced to quantify the vulnerability of a region of circuit layout to hardware Trojan insertion. Metrics enabled to quantitatively evaluate the vulnerability of each region of layout to various types of hardware Trojans. The results have indicated the considerable vulnerability of circuit layout to hardware Trojans resistant to delay-based, power-based, and multiparameter-based Trojan detection techniques. The analysis made is possible an in-detailed analysis of each gate and net of circuit, considering its placement and routing. Such an analysis may provide guidance for Trojan prevention during circuit development and Trojan detection after fabrication. The proposed vulnerability flow mainly considers hardware Trojan impacts on power and performance characteristics. However, a hardware Trojan may affect other characteristics such as electromagnetic or temperature as well. Therefore, a comprehensive layout vulnerability needs to consider various parameters to more accurately quantify the vulnerability of a circuit layout to hardware Trojans resilient to different hardware Trojan detection techniques.

References

1. ITC99 benchmarks. http://www.cad.polito.it/downloads/tools/itc99.html. Accessed 22 Jan 2018
2. Synopsys 90nm generic library for teaching ic design. http://www.synopsys.com/Community/UniversityProgram/Pages. Accessed 22 Jan 2018
3. L. Wang, C. Wu, X. Wen, *VLSI Test Principles and Architectures: Design for Testability*. The Morgan Kaufmann Series in Systems on Silicon (Morgan Kaufmann Publishers, San Francisco, 2006)
4. Ethernet 10GE MAC. http://opencores.org/project,xge_mac. Accessed 22 Jan 2018
5. Trust-HUB. https://www.trust-hub.org/. Accessed 22 Jan 2018
6. R. Karri, J. Rajendran, K. Rosenfeld, M. Tehranipoor, Trustworthy hardware: identifying and classifying hardware Trojans. Computer **43**(10), 39–46 (2010)

Chapter 7
Design Techniques for Hardware Trojans Prevention and Detection at the Layout Level

7.1 Hardware Trojan Prevention at the Layout Level

Hardware Trojan prevention at the layout level has mainly focused on design techniques that merely prevent hardware Trojan insertion in a circuit layout or facilitate hardware Trojan detection after design manufacturing. In this section some of major design techniques for hardware Trojan prevention are discussed.

7.1.1 Dummy Scan Flip-flop Insertion and Layout-Aware Scan Cell Reordering

Dummy scan flip-flop insertion and scan cell reordering techniques are two complementary design techniques that improve hardware Trojan detection resolution by increasing partial switching activity of hardware Trojan circuit and reducing overall circuit switching activity [1, 2]. The two techniques facilitate hardware Trojan detection based on power analysis. It is expected that hardware Trojan inputs are driven by signals with low switching activity to reduce the probability of full activation of hardware Trojan during circuit authentication. To increase hardware Trojan partial activation (i.e., increasing switching activity inside a hardware Trojan circuit), the dummy scan flip-flop insertion technique restitches signals with very low '1' or '0' probabilities [1]. Signals with very low '0' probabilities are restitched through the structure in Fig. 7.1a, and signals with very low '1' probabilities are restitched through the structure in Fig. 7.1b.

In Fig. 7.1, the signal TE is the test enable signal that distinguishes between the functional and test modes. The signal SI is the serial input signal that is either driven by another dummy scan flip-flop or by a primary input on which '1' and '0' are applied with the equal probability.

H. Salmani, *Trusted Digital Circuits*, https://doi.org/10.1007/978-3-319-79081-7_7

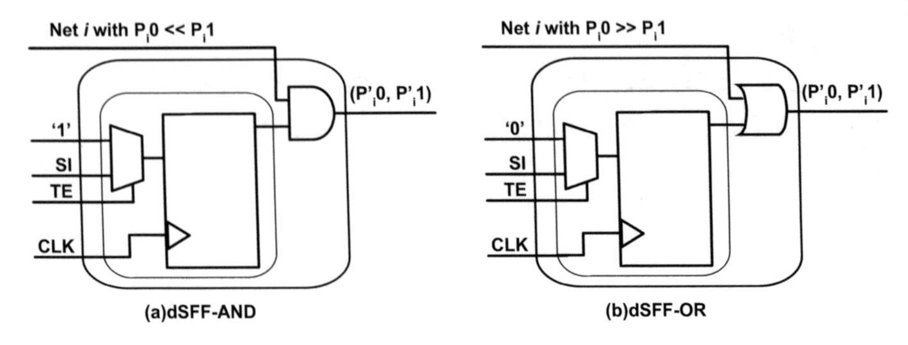

Fig. 7.1 The dummy flip-flop structures when (**a**) $P_i0 \ll P_i1$ and (**b**) $P_i0 \gg P_i1$ [1]

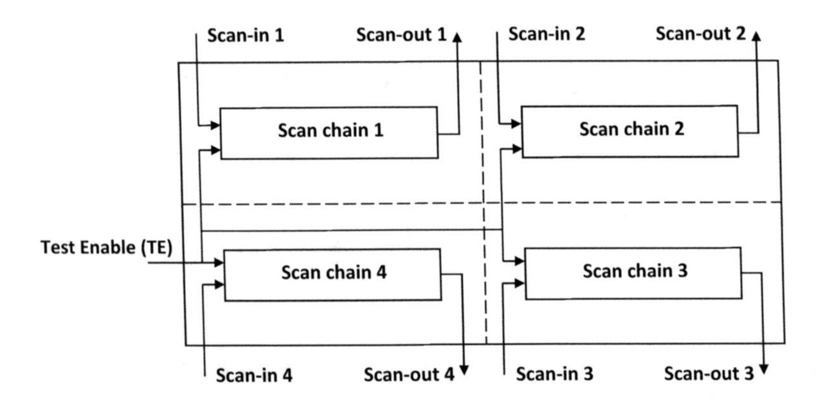

Fig. 7.2 Layout-aware scan-cell reordering concept [2]

If a signal (Net i in Fig. 7.1) has considerably low transition probability and the probability of '0' is significantly smaller than the probability of '1', Net i is passed through dSFF-AND in Fig. 7.1a. dSFF-AND increases the probability of '0' on net i in the authentication mode. In the functional mode, the functionality of Net i is remained intact. On the other hand, if Net i has considerably low transition probability and the probability of '1' is significantly smaller than the probability of '0', Net i is passed through dSFF-OR in Fig. 7.1b.

Hardware Trojan detection resolution depends on hardware Trojan activity directly and circuit activity reversely. The scan-cell reordering technique reorders scan cells based on their placement during physical design to reduce circuit switching activity by limiting it into a specific region [2]. This helps magnify hardware Trojan contribution to the total circuit transient power by increasing the ratio of hardware Trojan to circuit power consumption. Figure 7.2 shows the basic concept of the layout-aware scan-cell reordering. Assume that a design with four scan chains is divided into four regions. The method forms the scan chains such that scan cells placed in each selected region are connected to each other. It is to ensure that the scan chains have the same length, but that is not a requirement. The technique enables magnifying the hardware Trojan impact by increasing the ratio of hardware Trojan to circuit power consumption through maximizing switching in the target region (e.g., the region containing scan chain 4) while minimizing switching in all the other regions (1, 2, and 3).

7.1.2 Ring Oscillator Network

A ring oscillator is a chain of odd number of inverting components where the last component is driving the input of the first component. In its simplest form, the ring oscillator is made of inverters. While any inverter in the ring oscillator produces a period waveform, the period of this waveform is determined by the total delay of all inverters in the presence of supply voltage variations and process variations. The ring oscillator can serve as a power monitor as well since its frequency depends on supply voltage and any variation on power supply changes the frequency of the ring oscillator. As any switching activity in a hardware Trojan circuit consumes power, any unexpected change in the frequency of a ring oscillator can indicate the existence of the hardware Trojan circuit near to the ring oscillator. Based on this assumption, a ring oscillators network (RON) where each serves as a power monitor is distributed across the circuit layout such that each row of the circuit under authentication contains at least one inverter of an RO in the RON, as shown in Fig. 7.3.

This on-chip structure includes a Linear Feedback Shift Register (LFSR), one decoder, one multiplexer, and one counter. The LFSR supplies random functional patterns for the entire circuit during the signature generation and authentication processes; the same seed must be used for each golden circuit and each circuit under authentication. A decoder and multiplexer are used to select which ring oscillator is measured. The output of a selected RO provides the clock input of the counter that measures the cycle count of the selected ring oscillator over a specified duration.

Assuming that a set of hardware Trojan-free circuits exist, the RON for each circuit is characterized by obtaining the frequency of each ring oscillator of the RON under the same functional patterns. Any circuit under authentication also becomes characterized under the same functional patterns. Any considerable mismatch beyond the impact of process variations flags the hardware Trojan existence.

7.1.3 Trojan Prevention and Detection (TPAD) Technique

Using a combination of new design techniques and new memory technologies, the hardware Trojan Prevention and Detection (TPAD) technique is to detect hardware Trojans during integrated circuit testing and during system operation in the field [4]. TPAD assumes that a trusted register-transfer level circuit or system specification exists and system assembly is also trusted. Therefore, logic synthesis tools, physical

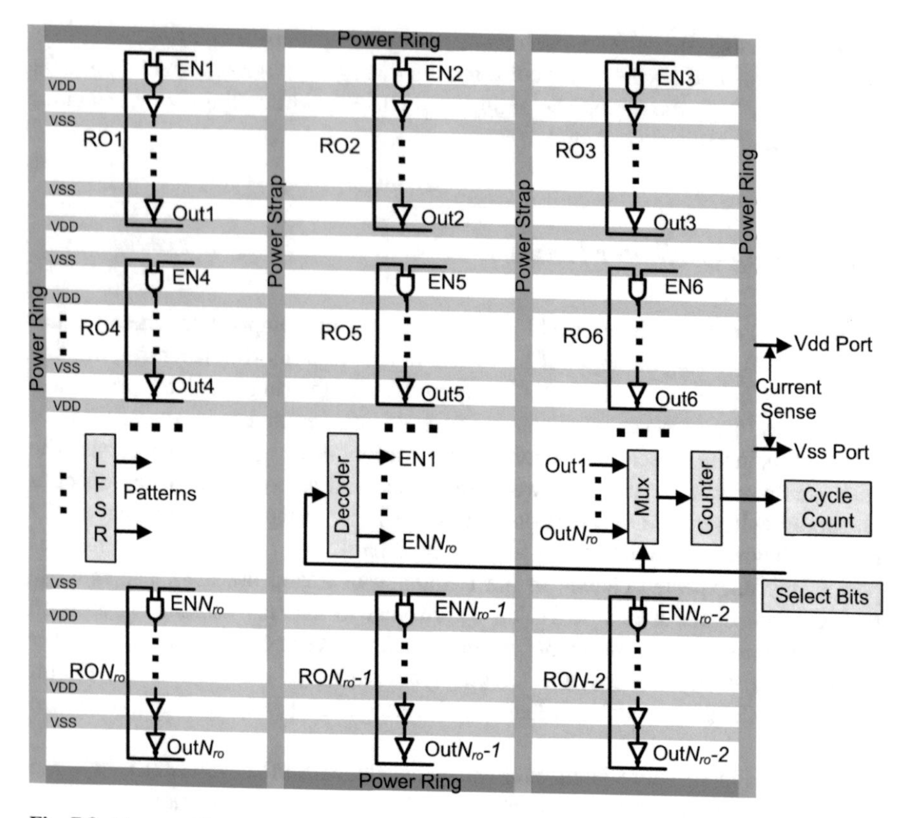

Fig. 7.3 The on-chip structure with each gate of the ring oscillators placed in a standard-cell row [3]

design tools, as well as IC fabrication are untrusted. Toward ultimate assurance, an entire circuit can be implemented using reconfigurable computing modules such as Filed Programmable Gate Arrays. However, the final design suffers significant area, performance, and power overhead. To mitigate the overhead, TPAD applies reconfigurability on only the critical portion of a circuit. A complex system may consist of several chips that communicate either directly or through an on-chip network. TPAD suggests encoding the communication between the chips to further prevent hardware Trojan attacks with the assumption the on-chip network is trusted by itself and a hardware Trojan exists in at least one of the chips. Figure 7.4 shows the implementation of TPAD at the system level, and Fig. 7.5 presents the detailed implementation of TPAD inside a chip.

Each chip in the system encodes its outputs and receives encoded inputs. Specifically, chip 1 outputs data and corresponding check bits, so chip 2 can use them to verify the data. Encoded error signals sent from each chip convey the state of all checkers within the chip. The error monitors then interpret these signals and determine if an attack has occurred. Each chip implemented using TPAD

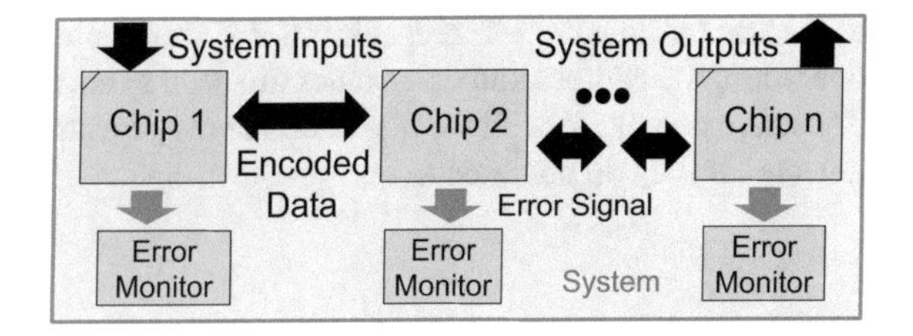

Fig. 7.4 Trusted system. Each chip is implemented with TPAD. Data communication between chips is encoded. Error monitors check encoded error signals and determine if an attack has occurred

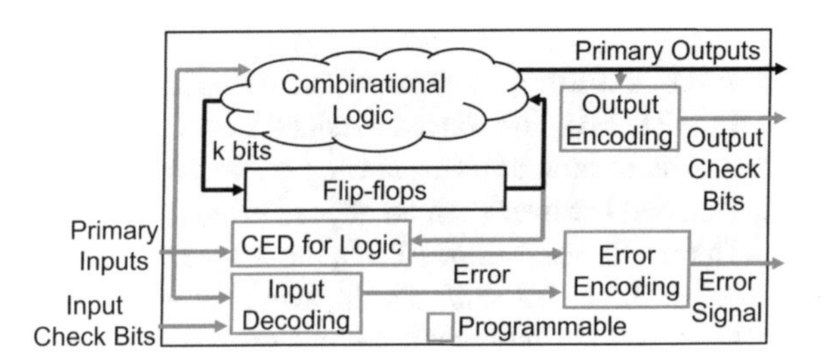

Fig. 7.5 TPAD architecture for each chip, including output encoding, input decoding, oncurrent error detection (CED) for logic, and error encoding [4]

includes four modules: (1) output encoding; (2) input encoding; (3) concurrent error detection-based hardware Trojan detection; and (4) error encoding.

Primary outputs of each chip are encoded using randomized parity encoding to generate output check bits accompanying the primary outputs. The output check bits are generated based on a randomized parity code word that is calculated during each clock cycle and depends on the previously generated check bits. Such a dependency over the time can reveal hardware Trojans that become activated at certain time and affect primary outputs of a chip. A separate encoding is used for each subsequent chip in the system that receives a different set of the primary outputs; the same encoding is used for chips receiving the same set of primary outputs. Concurrent error detection (CED) is used to protect the circuit logic as shown in Fig. 7.5. To deliver reasonable security, all combinational logic circuits should be first identified, and each combinational logic circuit becomes protected by a separate CED.

Error signals produced by CED checkers can be manipulated by a hardware Trojan if these signals are easily interpretable. To prevent such a attack, the error signals are protect by introducing randomness in the Error Encoding module in Fig. 7.5 by using a programmable linear feedback shift register (LFSR). The error signals then need to be examined by a trusted Error Monitor module (Fig. 7.4) to determine if an attack has occurred or not. The error monitor module includes LFSRs that are set with the same characteristic polynomial, seed, and clock as the LFSRs used in the error encoding module. To realize this error monitor, an older technology node from a trusted foundry may be used.

A hardware Trojan may target an on-chip memory module and interfere with its read and write operations or its address decoder. To detect such attacks, a randomized parity code is used to protect the memory module. In the write-through write operation, TPAD records the values on the address bus and the data bus, and calculates check bits based on these values to ensure that correct data are written to the correct location. The check bits are stored along with the data. In the read operation, the address and data values are used to calculate the expected check bits. These are compared with the check bits read out from the memory and an attack is detected if they do not match.

7.1.4 Infrastructure IP for Security (IIPS) Technique

In a similar effort to address security of complex designs, an Infrastructure IP for Security (IIPS) is introduced to address various security issues including scan-based attacks in cryptographic systems, reverse engineering, cloning and overproduction attacks, and hardware Trojan attacks in system-on-chip designs [5].

The block diagram of IIPS is provided in Fig. 7.6. It consists of a Master Finite State Machine that controls the working mode of IIPS, a Scan Chain Enabling FSM to provide individual control over activation of the scan chains in an SoC, and a clock control module to generate necessary clock and control signals for performing path delay-based hardware Trojan detection. Specifically targeting hardware Trojans, IIPS integrates the clock sweeping technique proposed in [6] for hardware Trojan detection through path delay measurement. IIPS provides a centralized infrastructure for characterization of path delays in individual IPs.

As tapping into any original signal would modify its delay characteristics, the clock sweeping technique collects the delay testing response over various clock frequencies, and then performs statistical analyses to identify possible hardware Trojans. IIPS makes the clock frequency tunable to support a fine-grained frequency

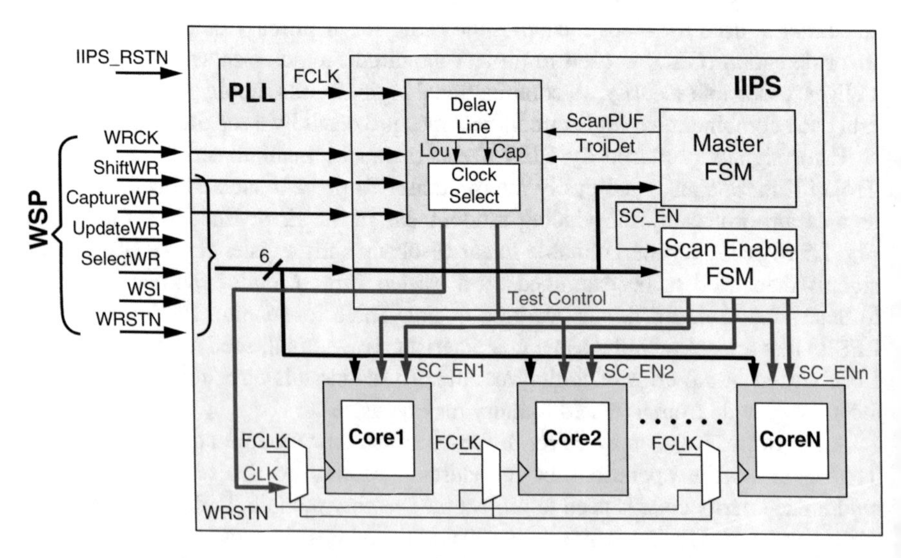

Fig. 7.6 Block diagram of the IIPS module showing an interconnection with other IP cores in an SoC using SoC boundary scan architecture [5]

sweeping over a wide range. Hardware Trojans might be mounted inside the functional cores or they could tamper with the interconnection among cores. IIPS has two modes for hardware Trojan detection: (1) INTEST_SCAN that allows inward facing test for a a specific core while other cores are set to BYPASS mode; (2) EXTEST that allows the regional analyses of interconnection that are driven by a specific core while other cores are being set to BYPASS mode.

7.2 Hardware Trojan Detection at the Layout Level

The primary inputs and outputs are *logical* access and observation points (channels) of a chip. However, there are other channels that can reflect computation activities inside the chip that are known as side channels. For example, analyzing power consumption of a chip through monitoring power/ground pins can reveal what computation is being performed at a time by knowing the detailed implementation inside the chip. Side channels can take various forms such as timing information, electromagnetic information, thermal information, and power information. The side-channel information is of great concern in the field of cryptography as it extremely facilitates determining secret keys stored inside the chip. Considering effectiveness of side-channel attacks on crypto chips, various hardware Trojan detection techniques have been proposed to capture hardware Trojan footprints in side-channel signals. In the following some of these techniques are being discussed.

7.2.1 The Current Integration Technique

The amount of current drawn by a hardware Trojan can be so small that it submerges into the envelope of noise and process variation effects, where it cannot be detected by measurement equipment. However, hardware Trojan detection resolution can greatly improve if the current is measured locally from multiple power pads. The more instances of switching on hardware Trojan inputs and inside hardware Trojan circuitry, the greater the hardware Trojan's power consumption.

Since small Trojans are expected to be inserted into chips to reduce the probability of detection, the local current impact could be more significant than the global current measured from power pins. A current integration methodology is presented in [7] which accumulates the impact of a hardware Trojan over the time while it is expected that the process variations impact is canceled out by integration. Figure 7.7 shows the current integration methodology for detecting hardware Trojans.

It is assumed that an untrusted manufacturer inserts some hardware Trojans into a selected number of chips. The exhaustive testing on a few randomly selected chips can help identify some golden chips (i.e., chips without hardware Trojans). After identifying the golden chips, an average current waveform is formed in

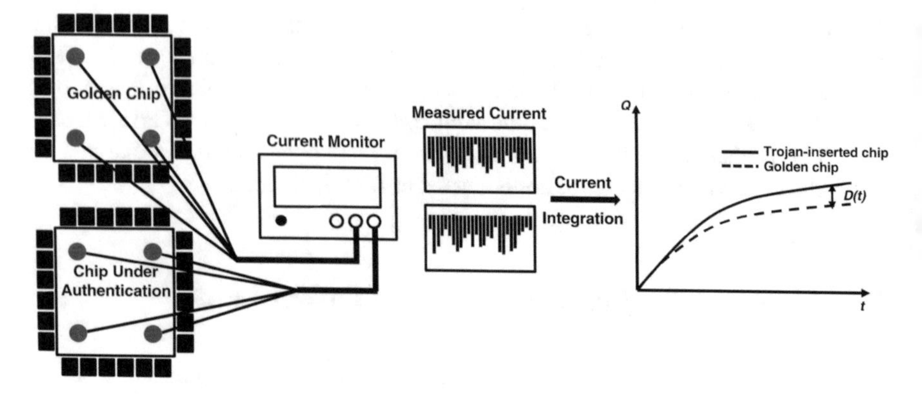

Fig. 7.7 Current integration method [7]

response to a test pattern set. Next, the test pattern set is applied to each circuit under authentication, and the current is measured locally via power pads or C4 bumps. Using this current integration method, the small current consumption difference between hardware Trojan-inserted and hardware Trojan-free circuits can be increased through the integration process. In the case of a hardware Trojan's existence in a chip, more current difference can be measured by applying more patterns to the chip, making hardware Trojan detection easier. When the current difference surpasses a pre-defined threshold, hardware Trojan existence is concluded and pattern application has stopped.

7.2.2 Delay-Based Hardware Trojan Detection Using Shadow Registers

Inserting hardware Trojans in only some of fabricate chips would demand additional masks that are expensive. Therefore, some work assumes that a hardware Trojan is inserted in all manufacture chips if an untrusted manufacturer wants to insert. Some techniques including [8] have been proposed to detect hardware Trojans based on delay analyses. As a hardware Trojan is expected to cause minimal delay deviations, a path selection scheme is proposed to select paths which maximize the additional delay induced by the hardware Trojan with respect to the nominal path delays and effects of process variations [8]. The path selection scheme targets timing paths having the smallest path delay values to maximize the impact of the hardware Trojan on each path's delay. Further, new logic and timing conditions are derived that must be satisfied to detect a hardware Trojan at a desired confidence level and at a minimum cost.

It is assumed that a hardware Trojan is tapped into one or more original signals of a circuit in a manner that changes the delay of the original signals

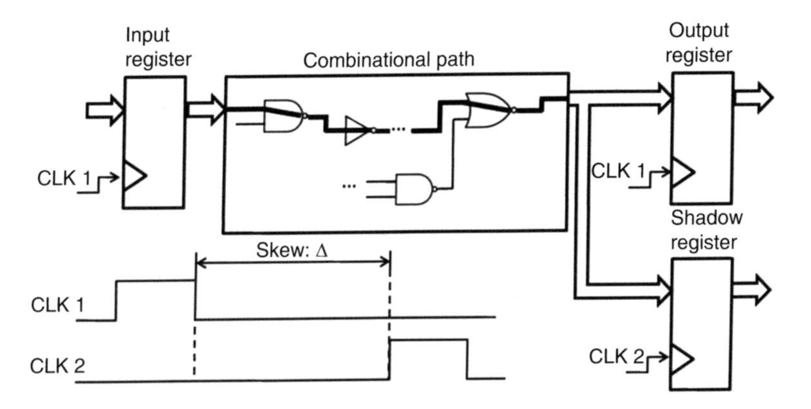

Fig. 7.8 Path delay measurement architecture [8]

even minimally. To measure small hardware Trojan-induced delay, the path delay measurement architecture in Fig. 7.8 is proposed. Taking advantage of the existing scan architecture on modern designs, the output of any target timing path is routed to a second register (the shadow register in Fig. 7.8). The shadow register is driven by a different clock which is being applied with the skew Δ relative to the circuit main clock. The approach does not require high-frequency clocks so that it avoids problems associated with excessive heat dissipation during testing of short delay paths. The total delay of a timing path P consists of three parameters: (1) nominal delay of path P, (2) the effect of process variations on the delay of P, and (3) extra delay induced by a hardware Trojan, if it exists. The effect of process variations follows random distribution, which is typically *bi-directional* around a mean. In contrast, extra delay induced by surrogate is *uni-directional* and always increases the total delay of path P. The proposed architecture is being applied on short timing paths as induced delay by a hardware Trojan constitutes a larger percentage of total delay of a short timing path compared to that of a long timing path.

While the technique assumes that it exists in all fabricated chips if a hardware Trojan is inserted by an untrusted manufacturer, the following question is what is the minimum number of fabricated chips sufficient to provide the correct disposition of the fabricated chips – either hardware Trojan-free or hardware Trojan-inserted – with low probability of decision error. A nonparametric test is proposed to identify the existence of the target hardware Trojan using the likelihood-ratio test.

7.2.3 Temperature-Based Hardware Trojan Detection

As modern and sophisticated designs utilize various sensors including thermal sensors for dynamic thermal management, some techniques including [9] have been proposed for online detection of deviations in power/thermal profiles caused

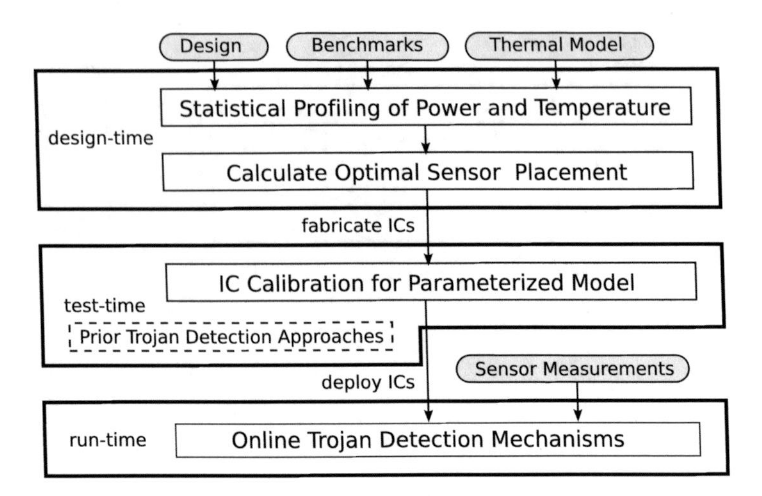

Fig. 7.9 Phases of the temperature tracking approach [9]

by hardware Trojan activation. The proposed framework for temperature-based hardware Trojan detection consists of three major phases: design-time, test-time, and run-time phases (shown in Fig. 7.9). In the design phase, an IC's power/thermal dynamics is statistically characterized. The test-time phase is used to calibrate each manufactured design due to fabrication variations. The run-time phase integrates the information from the previous phases with thermal sensor measurements to detect an activated hardware Trojan. Two mechanisms are suggested to detect hardware Trojan activation during run-time. The first is a local sensor-based approach that uses information from thermal sensors, statistical information provided by the test phase, and hypothesis testing. The second is a global approach that exploits correlation between sensors and maintains track of the IC's thermal profile using a Kalman filter (KF).

There is a strong correlation between temperature and power consumption. Therefore, any change in circuit power consumption due to any switching activity inside hardware Trojan circuit should also be reflected in circuit thermal profile. There are two major challenges to overcome: sensor infrastructure and noise, and fabrication variations. Firstly, temperature tracking highly depends on sensor placement, number of sensors, and sensor noise. Secondly, fabrication variations make it more challenging to track temperature as well as detect hardware Trojans as (1) It introduces larger uncertainty in the estimated thermal profile, and (2) It makes it difficult to distinguish between deviations in power/temperature due to manufacturing and hardware Trojan presence.

The above challenges are addressed in three phases (Fig. 7.9). At the design-time phase, assuming some hardware Trojan-free designs are available, the power/temperature profiles of hardware Trojan-free designs are obtained using the RC thermal model, benchmarks, and either state-of-the-art simulation tools or prototype ICs. Further, sensor placement is performed to optimize the temperature tracking/hardware Trojan detection. In the second phase, parameter calibration is performed. This is needed as fabrication variations cause manufactured designs that present different physical, electrical, and performance parameters from the nominal design so that fabrication variations make it more challenging to accurately detect hardware Trojans and track temperature. Calibration can be accomplished by applying test vectors to a manufactured design, measuring power consumption, and estimating probability distribution functions (pdfs) after fabrication. In the third phase, run-time, it is determined whether a circuit is hardware Trojan free/inactive or hardware Trojan inserted.

Two mechanisms are proposed to answer the above question. The first is a local sensor-based approach that uses a hypothesis testing (HT) framework. The second is a global approach that exploits correlation between sensors and maintains track of the design's thermal profile with a Kalman filter. In the hypothesis testing (HT) approach, one assumes that $S[k]$, one sensor measurement at timestep k, can only come from one of two pdfs: S_0 or S_1 which corresponds to temperature in hardware Trojan-free/Trojan-inactive and hardware Trojan-active designs respectively. The correct state is selected as the one with the highest probability of occurrence given $S[k]$ (i.e., argmax $Pr(H_x|S[k]), x \in \{0, 1\}$) where

$$\begin{cases} H_0 & \textit{The state is hardware Trojan}-\textit{free or hardware Trojan}-\textit{inactive} \\ H_1 & \textit{The state is hardware Trojan}-\textit{active} \end{cases}$$

In Kalman filter-based approach, the correlation between sensors is exploited and a Kalman filter is used to dynamically track the system's thermal profile at run-time. An autocorrelation-based metric then decides between hypotheses H_0 and H_1 (Trojan-free/Trojan-inactive and Trojan-active). The Kalman filter relies on (1) a state-space equation to model the random dynamics of the state being estimated and (2) a measurement equation to relate measurements with the state being estimated. The state-space equation for temperature tracking is the discrete form RC thermal model equation as following

$$\vec{T}[k] = \mathbf{A}\vec{T}[k-1] + \mathbf{B}\vec{P}[k-1]$$

The above equation assumes that the current thermal state $\vec{T}[k]$ depends on the previous thermal state $\vec{T}[k-1]$ and also local power dissipation $\vec{P}[k-1]$. Due to variations in the voltage supply noise, system workload, etc., the power $\vec{P}$ is random at each timestep and $\vec{T}[k]$ cannot be precisely computed with the state-space model alone. To improve the estimate, the Kalman filter uses measurements collected by thermal sensors and the following measurement model

$$\vec{S}[k] = \mathbf{H}\vec{T}[k] + \vec{v}[k]$$

where $\vec{S}[k]$ is a vector of sensor measurements at timestep k; $\mathbf{H}$ is a transformation matrix based on the sensor placement; and $\vec{v}[k]$ is a Gaussian random vector with zero mean and known covariance R representing measurement noise. The Kalman filter estimates the thermal state of a chip and it recursively performs predict and update based on new sensor measurement after each estimation. While the Kalman filter can accurately track temperature, a rule is needed to decide on the correct state (H_0 or H_1). The suggested decision rule is based on the Kalman filter residual and uses the autocorrelation function of the residual process. In the Kalman filter, residual $\vec{e}[k]$ indicates the difference between the predicated temperature and actual thermal sensor measurements. If $\vec{e}[k]$ is small (large), the predicted and actual temperature measurements agree (disagree) on the thermal state. Assuming the state-space model/parameters and the sensor noise covariance are reasonably accurate, the autocorrelation of the residual should be close to zero on average. When a hardware Trojan becomes activated, the state-space model (which does not account for the power of an active Trojan) becomes less accurate and should cause the autocorrelation to diverge from zero.

7.2.4 Circuit Layout Reverse Engineering for Hardware Trojan Detection

For hardware Trojan detection at the layout level, many works rely on the existence of a hardware Trojan free design obtained through reverse engineering. It should be noted that performing reverse engineering is very expensive and takes lots of time and consumes intensive manual effort. Meantime it is also very error prone, and there is a need for a robust reverse engineering to identify hardware Trojan free designs [10].

Three types of hardware Trojans at the layout level are being considered in [10]: (1) Hardware Trojan Addition (TA): These hardware Trojans add transistors, gates, and interconnects into the original layout. (2) Hardware Trojan Deletion (TD): These hardware Trojans delete transistors, gates, and interconnects from the original layout. (3) Hardware Trojan Parametric (TP): These hardware Trojans perform physical changes to the transistors, gates, and interconnects of the original layout.

The goal is to determine a provided chip by an untrusted manufacturer is hardware Trojan inserted or hardware Trojan free. This problem can be viewed as a classification problem described as following. Assume there are two classes of objects denoted by C_0 and C_1. Let each object (e.g., an untrusted chip) be represented by a feature vector $\mathbf{x} = (x_1, \ldots, x_n)$, where x_i denotes the ith feature, $x_i \in R$, and n denotes the number of features. Given an untrusted chip A, the problem is to determine the correct class of the chip A: hardware Trojan free or hardware Trojan inserted that contains TA, TD, and TP. It is assumed that a batch of chips manufactured by the same foundry will be either all hardware Trojan free or all hardware Trojan-inserted.

As a hardware Trojan is inserted by an untrusted foundry, it is assumed that the original physical layout (referred to as the golden layout) for all the layers of the

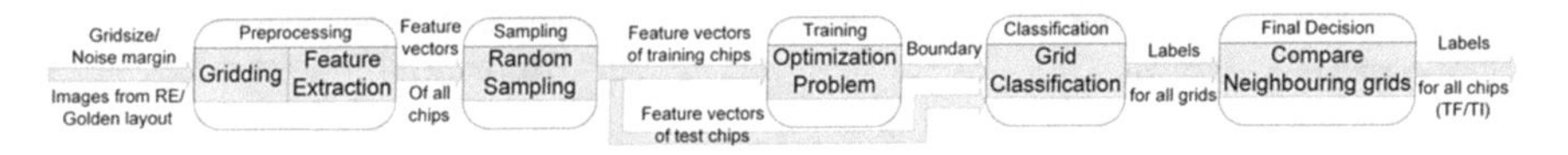

Fig. 7.10 The block diagram of SVM-based hardware Trojan detection approach [10]

chip exists. Figure 7.10 presents the overall approach for hardware Trojan detection using one-class support vector machine (SVM). SVM is a classification approach striving to classify some objects based on some of the objects' features. SVM is first trained on training samples to obtain a classifier and then uses this classifier to classify a given object.

The golden layout is taken as an input, N chips to classify, and parameter values for grid size, noise margin d_{nm}, etc. The N chips undergo some steps of reverse engineering that give images for all layers of all N chips. In the following, the images are divided as the images for each layer of all N chips into nonoverlapping grids. Breaking the images into smaller grids enables parallelized or distributed processing for each grid. Afterward, features are extracted for all N chips for each grid in each layer. The classifier is then trained and a decision boundary for each layer is obtained using a subset of the chips. After training, the grids in the each layer of all the N chips are classified as hardware Trojan free or hardware Trojan inserted based on the ν-SVM decision boundaries of each layer. Finally, each chip is labeled based on these grid classifications.

Features are key in labeling chips. Five features are used and they are determined based on area and centroid differences between the reverse-engineering and golden layouts, referring to Fig. 7.11. The first three features are obtained by calculating the intersection of different areas between the golden layout and reverse-engineered layout. They are given by the following equations:

$$f_1 = \frac{A(Y \cap Z_{in})}{A(Z_{in})}$$

$$f_2 = 1 - \frac{A(Y \cap \overline{Z}_{out})}{A(Y)}$$

$$f_3 = 1 - \frac{A(Y \cap \overline{Z}_{out})}{A(\overline{Z}_{out})}$$

Equations for the two centroid difference features are as follows:

$$f_4 = \frac{|CX(Z) - CX(Y)|}{\text{grid's length}}$$

$$f_5 = \frac{|CY(Z) - CY(Y)|}{\text{grid's height}}$$

where $CX(Z)$ and $CY(Z)$ denote the x and y coordinates of Zs centroid.

The final labeling for each chip is decided by looking at the number of grids classified as hardware Trojan inserted and their location. A chip is not classified as hardware Trojan inserted if it has only a few sparse hardware Trojan inserted grids. Rather, it is assumed that in order for the chip to be classified as hardware Trojan

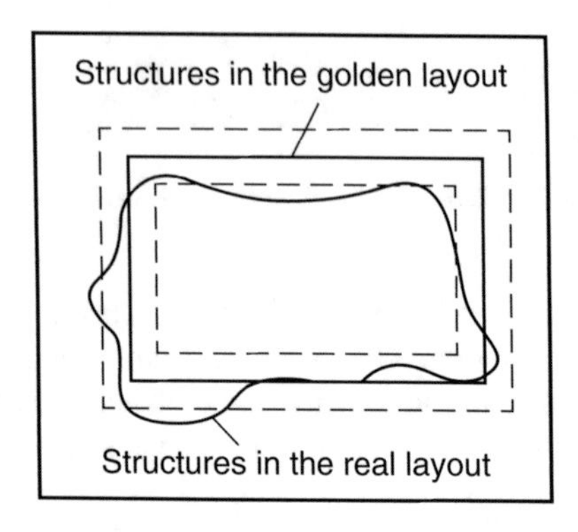

Fig. 7.11 Golden layout and SEM image of real layout within one grid. Solid rectangle and curve denote Z and Y, respectively, while rectangles in dashed line are Z_{in} and Z_{out} [10]

inserted there must be at least n neighboring hardware Trojan inserted classified grids in the chip. Neighboring grids can be defined within layers (horizontally adjacent grids) and between neighboring IC layers (vertically adjacent grids).

7.3 Conclusions

This chapter studied some of existing techniques that aim at developing design for hardware trust techniques and facilitating hardware Trojan detection after design manufacturing. These techniques are mainly to improve hardware Trojan detection based on analyzing side-channel signals such as power, delay, and temperature. Some techniques benefit from various statistical analyses to determine the confidence level in their conclusion whether a manufactured design is hardware Trojan inserted or not. They have presented some degrees of success considering their false positive and false negative analyses. However, some of outstanding challenges are the need for hardware Trojan free circuit as a reference model, and their considerable time requirement and cost. As a result, there is still a need for effective techniques that can address hardware Trojan detection and prevention at the layout level.

References

1. H. Salmani, M. Tehranipoor, J. Plusquellic, A novel technique for improving hardware Trojan detection and reducing Trojan activation time. IEEE Trans. Very Large Scale Integr. Syst. **20**(1), 112–125 (2012)
2. H. Salmani, M. Tehranipoor, Layout-aware switching activity localization to enhance hardware Trojan detection. IEEE Trans. Inf. Forensics Secur. **7**(1), 76–87 (2012)
3. X. Zhang, A. Ferraiuolo, M. Tehranipoor, Detection of trojans using a combined ring oscillator network and off-chip transient power analysis. J. Emerg. Technol. Comput. Syst. **9**(3), 25:1–25:20 (2013)
4. T.F. Wu, K. Ganesan, Y.A. Hu, H.S.P. Wong, S. Wong, S. Mitra, TPAD: hardware Trojan prevention and detection for trusted integrated circuits. IEEE Trans. Comput. Aided Des. Integr. Circuits Syst. **35**(4), 521–534 (2016)

5. X. Wang, Y. Zheng, A. Basak, S. Bhunia, IIPS: infrastructure IP for secure SoC design. IEEE Trans. Comput. **64**(8), 2226–2238 (2015)
6. K. Xiao, X. Zhang, M. Tehranipoor, A clock sweeping technique for detecting hardware Trojans impacting circuits delay. IEEE Des. Test **30**(2), 26–34 (2013)
7. X. Wang, H. Salmani, M. Tehranipoor, J. Plusquellic, Hardware trojan detection and isolation using current integration and localized current analysis, in *2008 IEEE International Symposium on Defect and Fault Tolerance of VLSI Systems* (2008), pp. 87–95
8. B. Cha, S.K. Gupta, Trojan detection via delay measurements: a new approach to select paths and vectors to maximize effectiveness and minimize cost, in *2013 Design, Automation Test in Europe Conference Exhibition (DATE)* (2013), pp. 1265–1270
9. D. Forte, C. Bao, A. Srivastava, Temperature tracking: an innovative run-time approach for hardware Trojan detection, in *2013 IEEE/ACM International Conference on Computer-Aided Design (ICCAD)* (2013), pp. 532–539
10. C. Bao, D. Forte, A. Srivastava, On reverse engineering-based hardware Trojan detection. IEEE Trans. Comput. Aided Des. Integr. Circuits Syst. **35**(1), 49–57 (2016)

Chapter 8
Trusted Testing Techniques for Hardware Trojan Detection

8.1 Fault Simulation-Based Test Pattern Generation

Given a specific trigger condition, it is not possible to target an arbitrary signal of the circuit, as the effect of the induced logic malfunction by a hardware Trojan might not become propagated to an observation point. Based on this observation, a fault simulation-based framework is proposed that enumerates the feasible hardware Trojan payload signals for a specific triggering condition [1]. The framework benefits from a genetic algorithm (GA)- based Automatic Test Pattern Generation (ATPG) technique, enhanced by automated solution to an associated Boolean Satisfiability problem. The inherent parallelism of GA enables relatively rapid exploration of a search space; therefore, GA can deliver reasonably good test coverage over the fault list very quickly. Meanwhile, GA does not guarantee the detection of all possible faults, specially hard-to-detect faults. On the other hand, SAT-based test generation has been found to be remarkably useful for hard-to-detect faults. It has an interesting feature that it can declare whether a fault is untestable or not.

In theory any combination of rare signals can be used as a hardware Trojan trigger. However, many of these trigger conditions are not actually satisfiable, and thus cannot constitute a feasible trigger. Assuming small combinational and sequential hardware Trojans, in practice, most of the easy-to-excite trigger conditions as well as a significant number of hard-to-excite trigger conditions can be detected by the GA within reasonable execution time. The remaining unresolved trigger patterns are fed to the SAT tool; if any of these trigger conditions is feasible, then SAT returns the corresponding test vector. Otherwise, the pattern will be declared unsolvable by the SAT tool itself.

Figure 8.1 presents the test generation and evaluation flow for hardware Trojan detection. It is assumed that hardware Trojans are small (the number of inputs for a hardware Trojan is up to four), and they flip the value of a payload signal upon activation. The probability analysis is performed to obtain the transition probability of signals in the circuit netlist. Assuming small hardware Trojans, random sampling

H. Salmani, *Trusted Digital Circuits*, https://doi.org/10.1007/978-3-319-79081-7_8

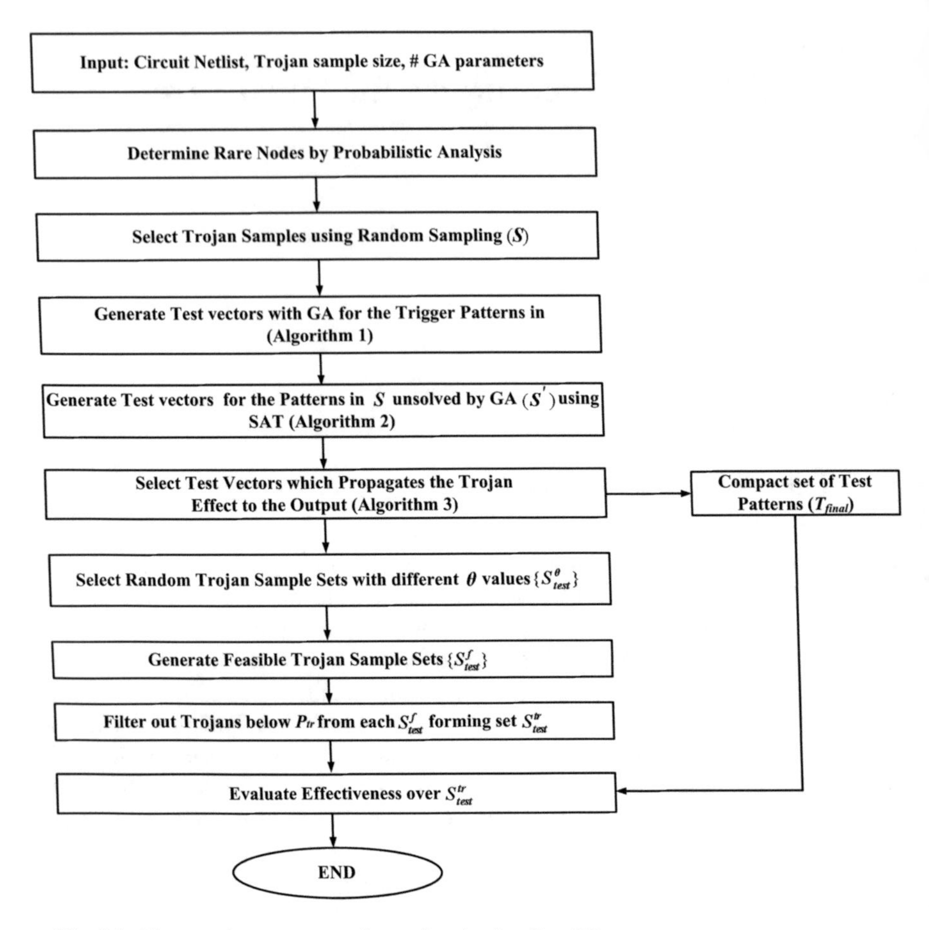

Fig. 8.1 The complete test generation and evaluation flow [1]

is performed by selecting a number of combination of rare signals as hardware Trojan trigger signals (S in Fig. 8.1). In the following, GA is used to generate test vectors for S in Algorithm 1 in Fig. 8.1. The initial test vector population for GA is generated by solving a small number of triggering conditions using SAT. Crossover and mutation operations are used to generate new test vectors. Two terminating conditions are used: (1) when the total number of distinct test vectors crosses a certain threshold value #T, or (2) if 1000 generations had been reached.

GA cannot guarantee that would be able to activate all the hard-to-trigger patterns. SAT can be used to tackle this issue (Algorithm 2 in Fig. 8.1). Considering S' as subset of S that contains trigger conditions that GA could not satisfy, SAT is solved for each $s \in S'$ to a test vector to satisfy the trigger condition s.

While GA and SAT strive to obtain a test vector for a triggering condition, this provides no guarantee regarding the propagation of hardware Trojan payload by the test vector to the primary output to cause a functional failure of the circuit.

Further, it is possible that GA and SAT report several test vectors for one triggering condition $\{t_i^s\}$. Algorithm 3 in Fig. 8.1 generates a 3-valued pseudo test vector (PTV) out of all patterns for a triggering condition. The pseudo test vector is obtained by crossmatching vectors and leaving bits with the constant value intact and replacing the others by don't care (X). At the next step, a three-value logic simulation of the circuit with the PTVs is performed, and the value of any signal in the fan-out cone of any triggering signal in a trigger vector is recorded. In the following, a stuck-at fault is defined over any signal in the fan-out cone of the trigger vector based on the value observed during the logic simulation. The list of faults F_S and $\{t_i^s\}$ is passed to fault simulator whose outputs are the set of faults that are detected ($F_{detected}^S \subseteq F_s$) as well as the corresponding test vectors which detect them. The rest of steps in Fig. 8.1 is performed to evaluate the effectiveness of generated test vectors.

The proposed test pattern scheme is based on the dual strengths of a genetic algorithm and Boolean satisfiability. The technique has presented good test coverage and compaction for small hardware Trojans while it needs more comprehensive analyses toward reasonably more realistic hardware Trojans.

8.2 Multiple Excitation of Rare Switching (MERS)

The effectiveness of side-channel signal analyses can be improved if there exit test vectors that highlight hardware Trojan contribution beyond large process variations. Multiple Excitation of Rare Switching (MERS) is a side-channel-aware test generation approach to improve hardware Trojan detection resolution based on power analyses [2]. An effective side-channel analysis in detecting hardware Trojans requires test vectors that (1) Maximize switching activities inside a hardware Trojan circuitry; (2) Minimize switching activities in the rest the circuit so that the relative switching effect is maximized. Figure 8.2 presents the Test generation framework for side-channel analysis-based hardware Trojan detection based on MERS [2].

First, logic simulation is performed using provided random test patterns to identify internal nodes with low probability that will be considered as rare nodes (R). In the following, it is counted how many rare nodes take their rare values (R_V) with each random pattern. The test patterns are then sorted decreasingly based on their R_V, meaning that the vector with ability to activate the most number of rare nodes is ranked first. Next, a rare switching counter (S_I), with initial value 0, is considered for each rare signal. Then the test patterns are applied, and S_I for each rare signal is determined. The goal is to have a rare signal obtain its rare value N (a predefined value). It is possible that there remain rare signals whose $S_I < N$ after applying the test patterns. Targeting only such rare signals, each test pattern is mutated by changing one bit and then logic simulation is performed. If the new test pattern increase any S_I, the test pattern is kept. This will be repeated till S_I for all rare signals becomes greater than N.

The obtained test patterns can increase switching activity of rare signals, but they may also create considerable switching activity over the entire circuit. As a result,

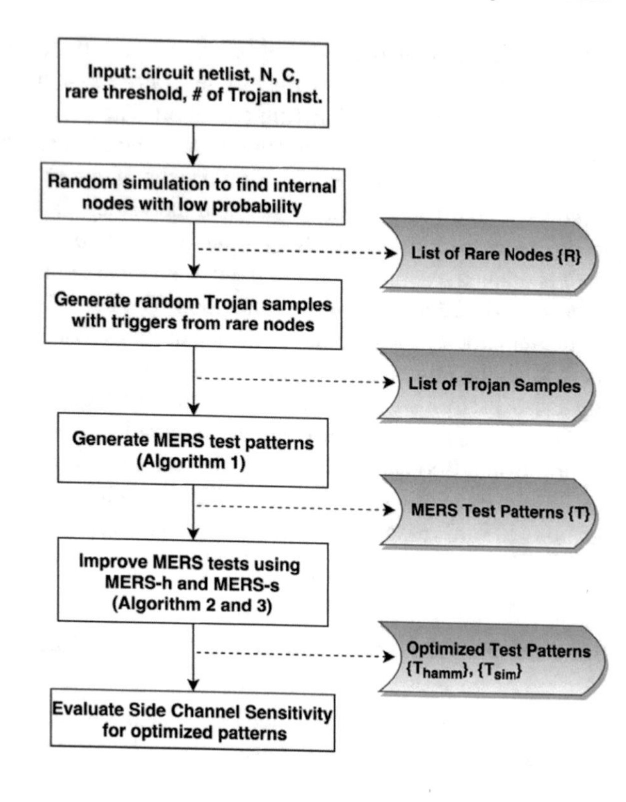

Fig. 8.2 Test generation framework for side-channel analysis-based hardware Trojan detection [2]

the hardware Trojan detection resolution will not increase significantly. To address the issue, two approaches are suggested: (1) A heuristic approach based on hamming distance of test vectors, which can reduce the total switching (2) A simulation-based technique strives to balance the rare switching and the total switching If two consecutive input vectors have the same values in most bits, it is very possible that the internal nodes will also have a lot of values in common. From a hardware Trojan detection perspective, the second vector would not generate significant switching activity over the entire circuit. Hamming distance between two vectors is used to represent the similarity. Then test patterns in the testset are reordered based on Hamming distance. The algorithm is a greedy approach to explore all candidate vectors and take the best one in terms of Hamming distance. MERS originally tries to increase switching of rare signals with goal of increasing switching activity inside a hardware Trojan circuitry. However, this objective is vague as the location and type of the hardware Trojan is unknown. A suggested solution is to consider rare switching between two vectors. To consider both switching activity inside a hardware Trojan circuitry and over the entire circuit, a new objective is defined as

$$maximiz\ (w_1 \times RareSwitch - w_2 \times TotalSwitch)$$

where the ratio of w_1 over w_2 is defined as C which is an input of the framework in Fig. 8.2. Using logic simulation, test patterns are reordered such that the objective becomes maximized.

8.3 Sustained Test Vector Methodology

To improve hardware Trojan detection based on power analyses, a sustained vector methodology is proposed at [3]. This methodology suggests repeating each test vector multiple times at the input of both the genuine and the hardware Trojan circuits that ensure the reduction of extraneous toggles within the genuine circuit. In the following, regions with wide variations in the power behavior are analyzed to isolate the infected gate(s). The proposed approach has two major steps. The first step (called Toggle Minimization) aims to detect the presence of a hardware Trojan while the second step (called Infected Region Isolation) tries to isolate the region within the circuit that may contain it.

Circuit power consumption depends on signal switching activity in a circuit. Knowing that power consumption by a hardware Trojan circuit is very small compared to the rest of the original circuit, it is essential that the overall power consumption in the genuine circuit should be minimized. Circuit activity within the combinational frame of the circuit is induced in two ways: (1) with the changing inputs and (2) with the changing state. Among these, primary inputs are fully controllable but the state variables are not. In order to limit the switching activity within the circuit, the proposed technique restricts the input variations to an extent that the state variables are the only factor for inducing toggles. This is achievable by sustaining the same vector at the input pins over multiple clock cycles. Naturally this would help in minimizing the overall circuit activity and keep the power consumption of the overall circuit low, which is a primary objective of the technique. Based on experiments by the authors [3], a set of 1000 random input vectors is generated, each of which is sustained to a maximum of 25 cycles. For a vector V, after sustaining it k times ($k < 25$), if it is found that the circuit has reached a stable state where no further change in the state variables occurs, the next test vector is applied. Holding the same vector at its input helps the circuit to traverse a local state space, and changing and sustaining the input vector serve as a jump to explore some other regions of the state space.

After applying sustained test vectors on both a genuine circuit and a circuit under test, a differential power profile plot is obtained. The differential power profile plot is used to to identify region(s) of the circuit that are likely to be insertion points of a hardware Trojan. Upon identifying a variation on circuit power consumption beyond the impact of process variations, the corresponding sustained vector, along with time of occurrence, is used for further investigation.

8.4 Monte Carlo-Based Test Pattern Generation Method

For testing against hardware Trojan, normally a fixed large number of random patterns are applied. Therefore, it is not guaranteed the length of test vector set is long enough to be representative or whether it is already over testing. To address these issues, a Monte Carlo (MC)- based test pattern generation method for hardware Trojan detection is proposed at [4]. Figure 8.3 presents the Monte Carlo-based hardware Trojan detection flow [4].

First, a metric called power percentage difference (PPD) for detection decision is defined based on the signal percentage difference between a golden circuit and circuit under test (CUT). PPD is defined as

$$PPD = \frac{P_{CUT} - P_{Golden}}{P_{Golden}}$$

where P_{CUT} and P_{Golden} are the estimated power values of the circuit under test and the golden chip, respectively.

Based on the flow (Fig. 8.3), both the golden circuit and circuit under test are first subjected to random patterns and the target signal, in this case, the total power dissipation, is estimated. After test vectors application, the logical values at primary outputs between the circuit under test and the golden circuit are compared. If there is a mismatch, it indicates hardware Trojan existence. Otherwise, PPD is calculated between the golden chip and the circuit under test. The new PPD is combined with the previous PPDs. New mean and standard deviation of the PPD values are calculated. The new PPD value is checked to see whether it satisfies the defined error level. For a desired confidence level $1 - \alpha$ with a given percentage error ϵ, the stopping criterion of simulations can be written as

$$\frac{t_{\alpha/2} \times S_T}{\mu_T \times \sqrt{N}} < \epsilon$$

where N is the number of iterations, $t_{\alpha/2}$ is obtained from the t distribution, μ_T the sample average, and S_T sample standard deviation.

If the error is beyond the error-tolerance level, the simulation will move to the next sample. Or else, the simulation terminates at the end of this sample. After that, the MC simulation results are then reported, including the number of samples and the average PPD value. Finally, the average PPD value is compared to the threshold, if it is less than the threshold, the circuit under test is deemed as hardware Trojan-free. Otherwise, it is considered as hardware Trojan-inserted.

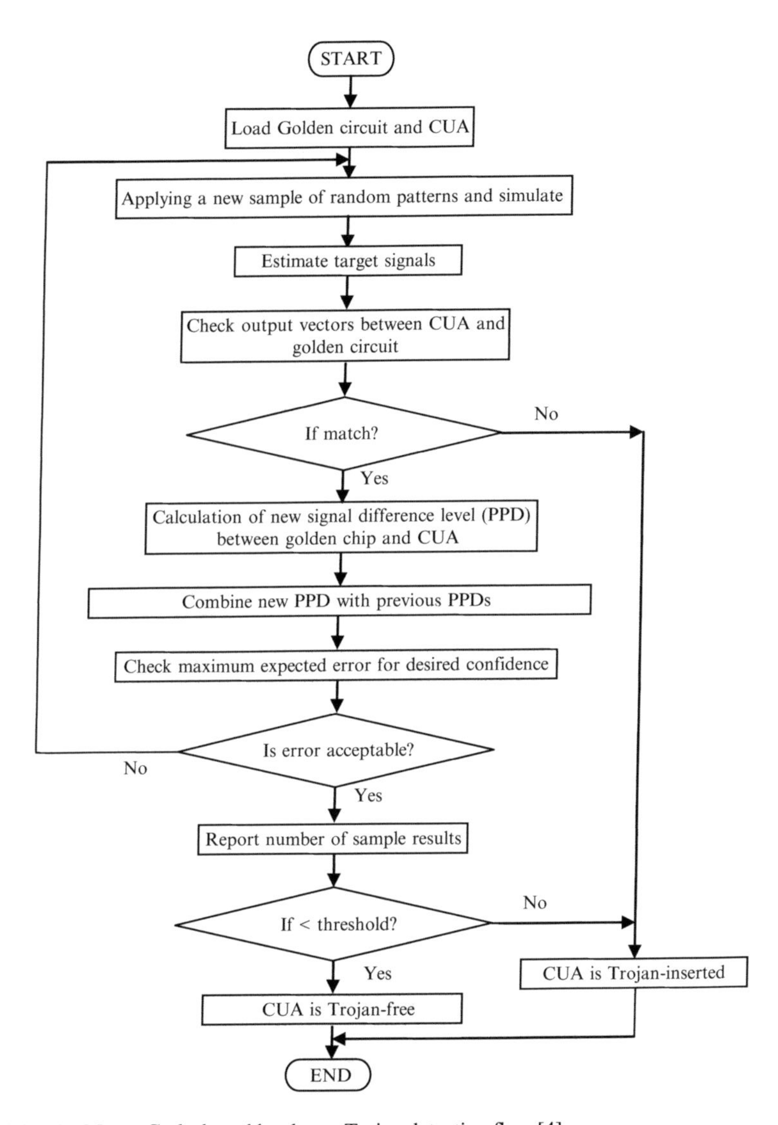

Fig. 8.3 The Monte Carlo-based hardware Trojan detection flow [4]

8.5 Test Pattern Generation Based on ATPG and Model Checking

Any hardware Trojan detection technique should not suffer from completeness and scalability. Further, it should feature efficient tests generation to activate the potential hidden hardware Trojan. A test generation approach is proposed at [5] to enable activating malicious functionality hidden in large sequential designs.

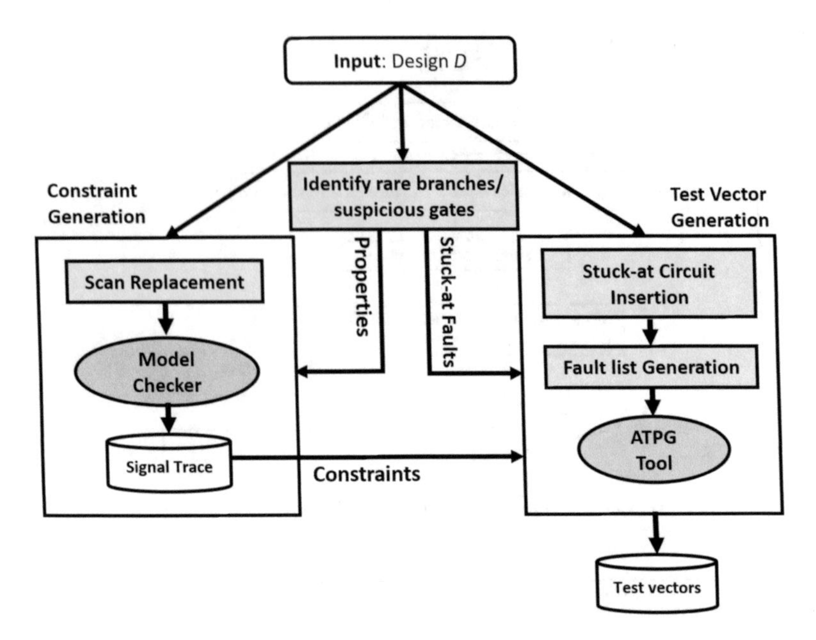

Fig. 8.4 The hardware Trojan detection flow using ATPG and model checking [5]

Automatic test pattern generation (ATPG) works well on full-scan designs, whereas model checking is suitable for logic blocks without scan chain. The proposed technique combines ATPG and model checking approaches for hardware Trojan detection. Model checking is used on a subset of non-scan elements and ATPG on scan elements to avoid common pitfalls of running the original design using any one of these techniques. Figure 8.4 presents the proposed hardware Trojan detection flow using ATPG and model checking [5]. The method identifies suspicious branches/gates which may be used as triggering conditions for hardware Trojans. In order to generate tests for rare nodes activation, scan replacement is done in the next step. Security properties are generated targeting the activation of equivalent signals/gates of rare nodes in the gate-level netlist. The scan replaced netlist as well as the security properties are then used by the model checker. A set of constraints is generated using the model checker to facilitate directed test generation using ATPG tool.

Initial analysis is performed at the register-transfer level (RTL) to isolate suspicious gates in the gate-level netlist. In a design at the register-transfer level, rare branches are defined as branches that are not covered after applying random test vectors up to millions of cycles. After identifying rare branches at RTL, mapping of the RTL branches to the gate-level netlist after design synthesis is performed. The circuit branches identified as rare are used in model checking property generation and ATPG stuck-at faults/node justification.

ATPG performance heavily depends on the scan architecture and the presence of non-scan sequential elements reduces its effectiveness. Model checking is used to generate traces transformed into constraint structures for the ATPG tool and facilitate test generation for rare nodes with non-scan flip-flops in their fan-in.

In the following, ATPG with N-detect testing is used to generate test vectors. The activation levels of all relevant internal signals from the suspicious node's fan-in cone and scan replacements are extracted from the trace and combined together with an ATPG primitive AND gate referred to as stuck-at circuit. The addition of the primitives is merely for test generation purposes and has no effect on the design functionality. Note that the propagation is not concerned as the goal is to generate switching activities on rare nodes. A stuck-at 0 fault is added to the tool's fault list corresponding to each stuck-at circuit. In the event that no test pattern can be generated for a rare node due to justification conflicts, nothing can be stated about the existence of a Trojan in the design.

8.6 Test Pattern Generation for Malicious Parametric Variations

A hardware Trojan may negatively impact design reliability through instantaneous increase of power consumption (*the power virus*) or extensive heat generation (*the temperature virus*) upon application of a specific test (input sequence) [6]. The power and temperature viruses can be targeted in a gate-level intellectual property using test vectors or processor intellectual properties using programs/benchmarks.

To detect the power virus at the gate level, the goal is to find a pair of test vectors that causes excessive power consumption by generating high transitions on circuit signals such that it exceeded a predefined threshold based on design specifications. An effective technique to know about the existence of such a pair is the cost-benefit analysis approach [6]. First gates are sorted by the number of fan-out number in decreasing order. The first gate (g) is selected, and it is tried to obtain a pair of test vectors that holds $g(x_1) \oplus g(x_2) = 1$ using backtracking and implication. This procedure is repeated for all gates in the circuit. While the procedure strives to obtain a pair of test vectors for each individual gate, the successive application of the pairs may impact their effectiveness in practice as early test vectors may restrict some other switching in following test vectors. Therefore, a cost-benefit analysis technique is proposed to facilitate the overall optimization process. The technique iteratively enables transitions in high fan-out gates while considering the trade-off between switching of new gates (benefit) and blocking of gate transitions in the future iterations (cost) due to switching of the currently selected one [7]. Figure 8.5 presents the cost-benefit analysis flow. The cost-benefit analysis selects the most beneficial gate in terms of the highest number of transitions. The gate is considered a tried gate and the obtained pair of test vectors is applied to the circuit, and the flow continues with an untried gate.

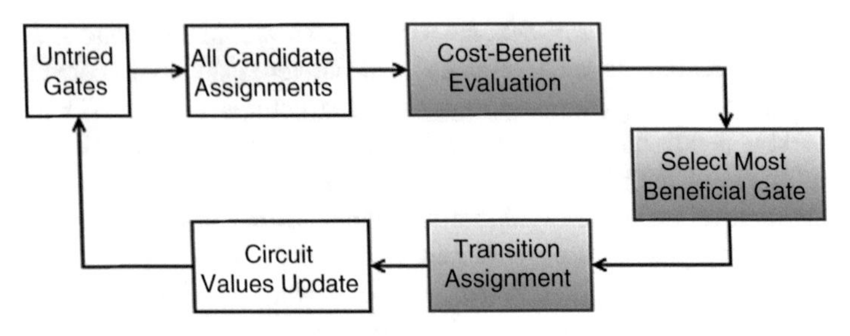

Fig. 8.5 The cost-benefit analysis flow [7]

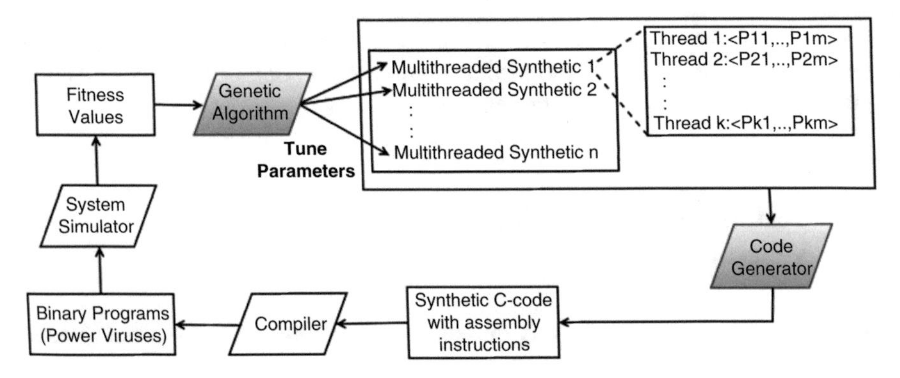

Fig. 8.6 A framework for multi-threaded power virus generation using a genetic algorithm [8]

A power virus can exist in complex circuits such as processors where the virus can cause excessive power consumption such that the embedded cooling system cannot manage the situation and the power management logic cannot stop the processor in time. Figure 8.6 presents a framework for multi-threaded power virus generation using a genetic algorithm [8]. A complex processor is made of several cores, tens of intellectual property modules that are connected through a network of chips. Several parameters need to be considered to find the extreme corners of power consumption. The parameters may include the number of threads, percent memory access to shared data, average branch predictability, integer instructions (ALU, mul, div), and mean of memory level parallelism. The framework in Fig. 8.6 uses the Genetic Algorithm to set these parameters. Afterward, a code generation is performed to generate a synthetic machine code based on set parameters. The processor executes the code using a simulator and it is determined whether fitness values for the genetic algorithm need an improvement. If needed the procedure is repeated and finally a program at the machine code level is obtained that evaluates the extreme power consumption of the processor. The temperature virus is a variation of the power virus where sustained high power consumption can produce peak temperature. Existing techniques for the power virus can be extended to obtain a set of successive test vectors that keep a design in its high power consumption.

8.7 Conclusions

There are two factors that determine hardware Trojan detection resolution after circuit manufacturing: switching activity of hardware Trojan circuit and switching activity of the original circuit. Some test pattern generation technique focus on one of factors and some consider both factors at the same time. While some techniques even benefit from sophisticated techniques such as SAT and the genetic algorithm, some outstanding challenging issues such as considerable false positive and negative rates, the need for a golden model, high reliance on sophisticated measurement tools for side-channel analyses, and considerable authentication time still remain. Therefore, there is a serious demand for effective and efficient test patterns for hardware Trojan detection in modern complex designs after their manufacturing.

References

1. S. Saha, R.S. Chakraborty, S.S. Nuthakki, Anshul, D. Mukhopadhyay, Improved test pattern generation for hardware Trojan detection using genetic algorithm and Boolean satisfiability. IACR Cryptology ePrint Archive 2015:1252 (2015)
2. R.S. Chakraborty, F. Wolff, S. Paul, C. Papachristou, S. Bhunia, *MERO: A Statistical Approach for Hardware Trojan Detection* (Springer, Berlin, 2009), pp. 396–410
3. M. Banga, M.S. Hsiao, A novel sustained vector technique for the detection of hardware Trojans, in *2009 22nd International Conference on VLSI Design* (2009), pp. 327–332
4. X. Mingfu, H. Aiqun, H. Yi, L. Guyue, Monte carlo based test pattern generation for hardware Trojan detection, in *2013 IEEE 11th International Conference on Dependable, Autonomic and Secure Computing* (2013), pp. 131–136
5. J. Cruz, F. Farahmandi, A. Ahmed, P. Mishra, Hardware Trojan detection using ATPG and model checking, in *International Conference on VLSI Design* (2018), pp. 1–6
6. Y. Huang, P, Mishra, Test generation for detection of malicious parametric variations, in *Hardware IP Security and Trust* (Springer, Cham, 2017)
7. H. Hajimiri, K. Rahmani, P. Mishra, Efficient peak power estimation using probabilistic cost-benefit analysis, in *2015 28th International Conference on VLSI Design* (2015), pp. 369–374
8. K. Ganesan, L.K. John, Maximum multicore power (MAMPO)—an automatic multithreaded synthetic power virus generation framework for multicore systems, in *2011 International Conference for High Performance Computing, Networking, Storage and Analysis (SC)* (2011), pp. 1–12

Chapter 9
Hardware Trojans in Analog and Mixed-Signal Integrated Circuits

9.1 Hardware Trojans Design in Analog and Mixed-Signal Integrated Circuits

While hardware Trojans in digital circuits are realized by inclusion of extra gates, hardware Trojans in AMS ICs do not necessarily incur any overhead. Hardware Trojans in AMS circuits can be realized by exploiting inherently available solutions due to feedback loops. A hardware Trojan in an AMS circuit can put the circuit under a specific condition to drive it to a malicious solution where the circuit behavior can be misused. The existence of analog and digital modules in one die makes it possible to design hardware Trojans that their activation in one domain could affect the execution of the other domain. In the following, some implemented hardware Trojans in AMS ICs are studied.

9.1.1 Hardware Trojans in Wireless Cryptographic ICs

Today applications such as internet of things bring together various analog (e.g., sensors, actuators, wireless communication modules) and digital modules (micro-processors). This trend considerably extends the attack surface as attackers without need for physical access can take the control of system. Authors in [1] have shown that how a hardware Trojan inserted in a cryptographic module can leak the secret key through a wireless transmitter. Figure 9.1 presents a wireless cryptographic IC that consists of a digital part and an analog module. The digital module is an advanced encryption standard (AES) core. The analog module is an ultra-wideband (UWB) transmitter (TX).

Two hardware Trojans are demonstrated. They do not have any trigger, and they are capable of transmitting the 128-bit AES key hidden in the wireless transmission power amplitude/frequency margins allowed for process variations while ensuring that the circuit continues to meet all of its functional specifications. The first

H. Salmani, *Trusted Digital Circuits*, https://doi.org/10.1007/978-3-319-79081-7_9

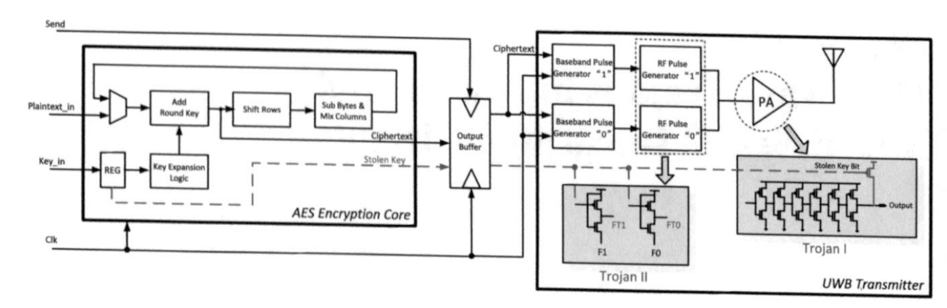

Fig. 9.1 Hardware Trojan modifications in digital and analog parts [1]

hardware Trojan (Trojan-I) is located in the power amplifier (PA), and the second hardware Trojan (Trojan-II) is located in the radio frequency (RF) pulse generators.

For Trojan-I, a pMOS transistor (shown in red in Fig. 9.1) is placed between VDD and the output of the PA of the UWB TX, and the stolen key bit is connected to the gate of this pMOS transistor. If the stolen key bit is ‘0’, the pMOS transistor is turned ON and draws a small additional current from the power supply to the output, thereby slightly increasing the transmission power. Conversely, when the stolen key bit is ‘1’, the pMOS transistor is turned OFF, so no additional current is drawn to the output, with the circuit, essentially, continuing to operate as in the Trojan-free case. For Trojan-II, two transistors are added at the input of each RF pulse generator (RF Pulse Generator ‘1’ and RF Pulse Generator ‘0’ in Fig. 9.1). When the stolen key bit is ‘0’, the pMOS transistor in Trojan-II is turned ON. By carefully crafting the sizes of added pMOS and nMOS transistors, it is possible to generate pulses with close but slightly higher frequencies than of original pulses. Figure 9.2 presents transmission power waveform of Trojan-II infested chip when the stolen key bit is ‘0’ and ‘1’ when transmitting a ciphertext bit of value ‘0’ or ‘1’ [1].

9.1.2 *Dynamic Analog Hardware Trojans*

An AMS circuit can have multiple modes of operations depending on some initial conditions, for example the initial values of storage elements. It has been shown that how it is possible to realize a hardware Trojan in the Wein bridge oscillator by manipulating some initial conditions as shown in Fig. 9.3 [2]. Figure 9.3a depicts the Wein bridge oscillator whose output (VOUT) to ground path passes two capacitors (C1 and C2) than can take different initial value. When the initial value of C1 and C2 is high, the circuit behaves as expected and oscillates as shown in Fig. 9.3b. However, if the initial value of C1 and C2 is low, the circuit does not oscillate as shown in Fig. 9.3c For the same Wein bridge oscillator, it is possible to have several dynamic oscillating modes under different initial conditions for C1 and C2. When the initial conditions added at C1 and C2 are the same, the circuit can oscillate with an amplitude (peak value) as 0.4 V. However, when the initial conditions added at

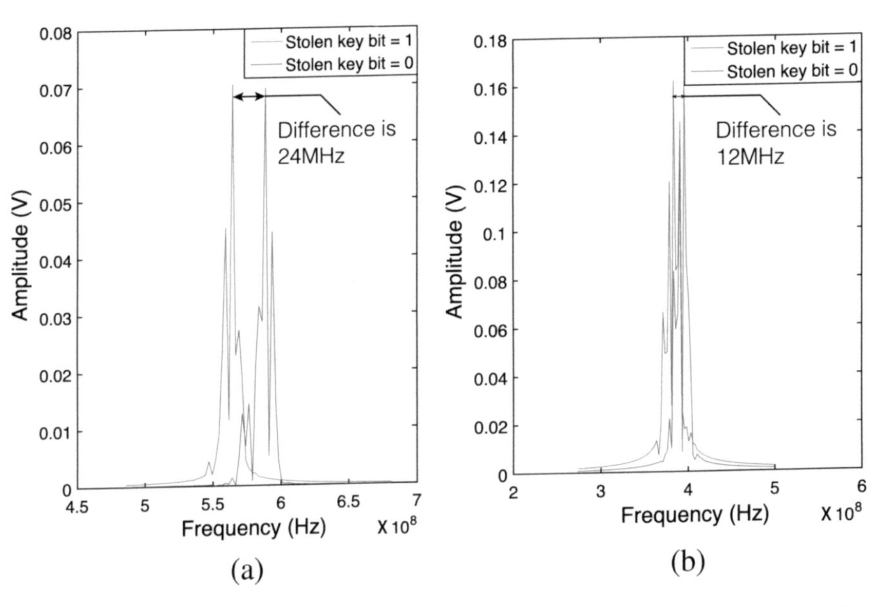

Fig. 9.2 Difference in transmission power waveform of Trojan-II infested chip when the stolen key bit is '0' and '1' (**a**) while transmitting a ciphertext bit of value '0', and (**b**) while transmitting a ciphertext bit of value '1' [1]

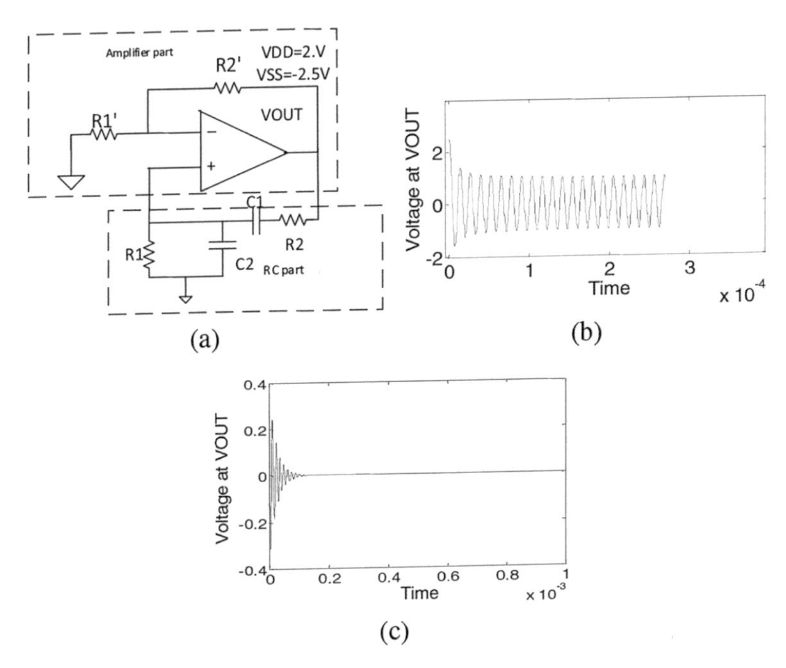

Fig. 9.3 (**a**) The Wein bridge oscillator, (**b**) a dynamic oscillating mode, and (**c**) a stable static mode [2]

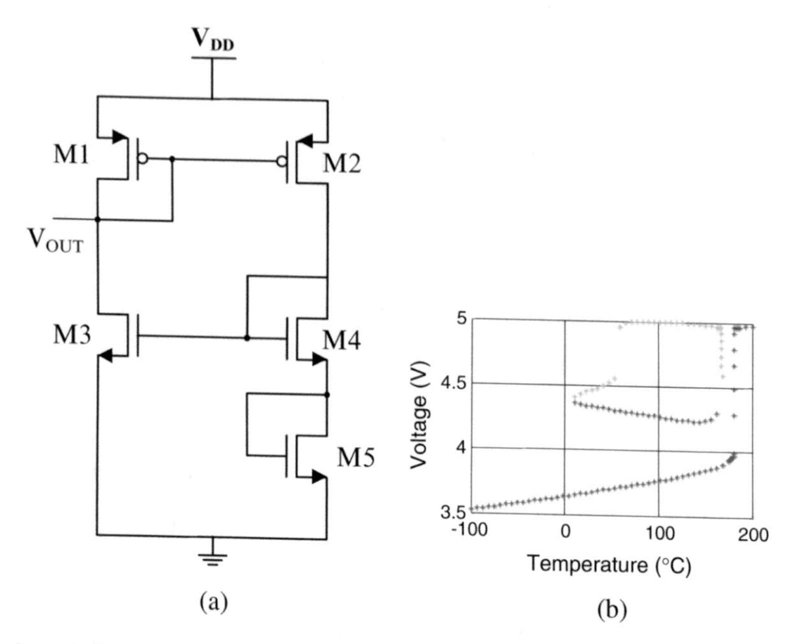

Fig. 9.4 (**a**) The Inverse Widlar Reference Generator, and (**b**) the temperature characteristics with three equilibrium points for $-4\,°\mathrm{C} < T < 173°\mathrm{C}$ when the width of transistors M1, M2, M3, M4, and M5 are 3 μm, 3 μm, 1.5 μm, 110 μm, and 6.5 μm, respectively [3]

the two capacitors are different, the circuit can oscillate in another mode, with a different amplitude 2.5 V, but a similar frequency.

Many basic analog circuits such as bias generators, voltage references, and temperature sensors use positive feedback or bootstrapping to generate outputs that have a low sensitivity to the supply voltage. Invariably these circuits incorporate positive feedback loops. These types of analog circuits often have two stable equilibrium points and a start-up circuit is typically added to remove the undesired equilibrium point. However, it can be very challenging to determine the need for a start-up circuit or ensure the effectiveness of a start-up circuit. Therefore, the redundant state may still exist that can harbor an analog hardware Trojan. Knowing the fact, insertion of an analog hardware Trojan in the inverse Widlar bias generator is studied in [3]. Figure 9.4a presents the Inverse Widlar Reference Generator circuit and Fig. 9.4b shows the voltage-temperature characteristics of the circuit. It clearly shows that there exist three equilibrium points when the temperature changes between $-4\,°\mathrm{C} < T < 173°\mathrm{C}$ (the isolation region). Therefore, it is possible to have three different output voltages at a specific temperature. For example at 100 °C, the voltage can be about 3.75, 4.25, or 5 V. If the normal operation is far from where the isolation region occurs and the isolation region occurs is even narrow, it is possible to realize a temperature triggering hardware Trojan even without an additional circuit overhead.

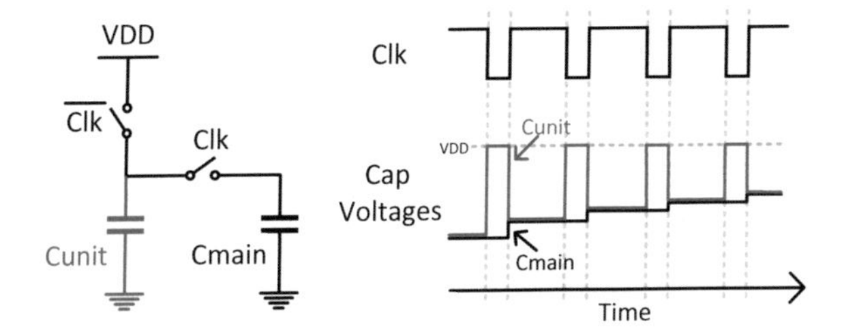

Fig. 9.5 Design concepts of analog trigger circuit based on capacitor charge sharing [4]

9.1.3 Hardware Trojan Trigger by Capacitor

It has shown that an untrusted manufacturer can leverage an analog circuit to realize stealthy hardware Trojans with inconsiderable footprint [4]. Figure 9.5 presents the basic concept of the analog hardware Trojan trigger. Figure 9.5a depicts in the insertion one extra capacitor (C_{unit}) that charges the level charge of an original capacitor (C_{main}). C_{unit} is charged when the level of a controlling signal (Clk) is low. The level of charge in C_{main} changes when Clk becomes '1'. Figure 9.5b shows how level charge in C_{main} raises from logic '0' to logic '1' after the burst alternation of the Clk signal. If C_{main} represents an active high reset signal of some state flip-flops, the state of circuit can change when C_{main} maliciously changes to logic '1' and it can cause catastrophic malfunctions. Authors in [4] used white spaces in an already placed and routed design to construct an analog hardware Trojan trigger that uses capacitors to siphon charge from nearby wires as they transition between digital values. They launched an attack on a victim flip-flop that holds the privilege bit in the OR1200 microprocessor. A wire that is controllable from software is selected and a malicious capacitor (indeed C_{unit}) is attached to it. The malicious capacitor serves as the analog hardware Trojan trigger and manipulates the set signal of the victim flip-flop.

9.1.4 Stealthy Dopant-Level Hardware Trojans

A stealthy hardware Trojan insertion by an untrusted manufacturer is proposed in [5]. This hardware Trojan is realized by changing the dopant polarity of a few existing transistors in the original circuit. The basic idea of this hardware Trojan is that a gate of the original design is modified by applying a different dopant polarity (changing dopant concentration) to specific parts of the active area of gate. These modifications change the behavior of the target gate in a predictable way. Figure 9.6 presents an example of dopant-level hardware Trojan implementation by manipulating an inverter that constantly produces the output VDD voltage (logic '1').

Figure 9.6a presents the original inverter whose upper part is a p-MOS transistor and lower part is an n-MOS transistor. The p-MOS transistor is realized by a N-

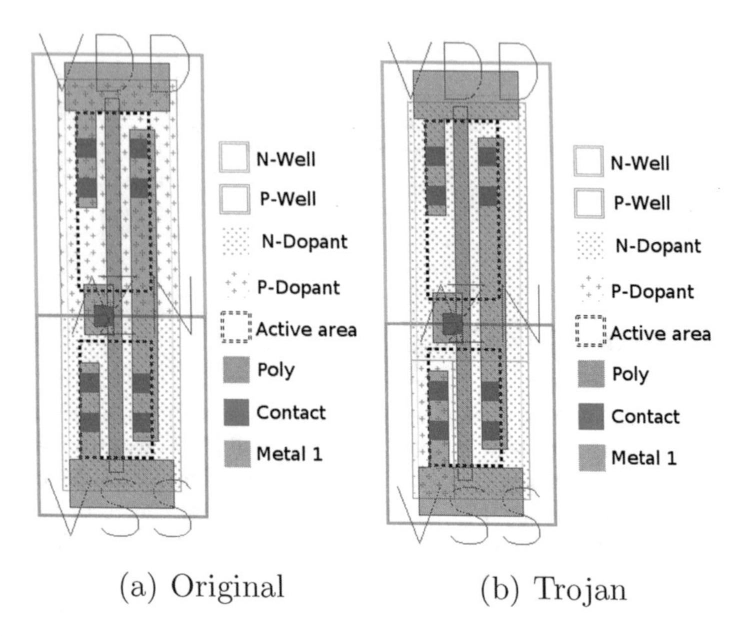

Fig. 9.6 (**a**) Figure of an unmodified inverter gate, and (**b**) figure of a Trojan inverter gate with a constant output of VDD [5]

well in which positive p-dopant for source and drain regions and the gate region are grown. On the other hand, the n-MOS transistor is realized by an P-well in which negative n-dopant for source and drain regions and the gate region are grown. To realize the dopant-level inverter hardware Trojan that constantly outputs VDD as shown in Fig. 9.6b, the positive p-dopant for source and drain and the active area of the p-MOS transistor is replaced by negative n-dopant. Consequently a permanent path created between VDD and the drain of p-MOS transistor. Further, the connection between the n-MOS transistor's drain contact and GND is constantly disabled by applying p-dopant to the source contact of the n-MOS transistor while leaving the drain contact untouched. This disconnects n-MOS transistor and GND independent from its gate input.

Showing the potential, the authors showed a hardware Trojan attack against a design derived from Intel's cryptographically secure random number generator design used in the Ivy Bridge processors, and a dopant hardware Trojan that allows attacking a side-channel resistant advanced encryption standard SBox implementation. Since the hardware is usually the root of trust in a system, even small malicious modifications of the hardware can be devastating to system security [5].

9.2 Hardware Trojans Prevention and Detection in Analog and Mixed-Signal Integrated Circuits

A hardware Trojan state in an AMS circuit may inherently exist, and an analog hardware Trojan mostly exists at the transistor level. The analog hardware Trojan can be even triggered by process, voltage, and temperature variations. For example, a circuit block may continuously work and increase its surrounding temperature, and the adjacent block in another power domain may enter into a hardware Trojan state when it starts up from the sleep mode [6]. In the following, some of the major hardware Trojan prevention and detection techniques in AMS ICs are studied.

9.2.1 Positive Feedback Loop Analyses

Positive feedback loops in analog circuits may create multiple operating points of which some can be hardware Trojan states. Circuit-level homotopy methods have been used to determine all operating points in the positive feedback loop circuit [6]. There exist two stable operating points of which only one is desired and one unstable operating point in a single positive feedback loop. Therefore, there exist two hardware Trojan state. AMS circuits are usually large and complex due to feedback structure and smart compensation techniques so that many hardware Trojan states may exist to be identified.

The proposed technique has two major steps: (1) identifying positive feedback loops in a circuit, and then (2) applying the homotopy methods to each positive feedback loops. To identify the positive feedback loops, the circuit is transformed to a directed dependency graph, and the positive feedback loops are identified based on the graph theory and the small signal analysis of the circuit. The positive feedback loops are broken and current or voltage source is inserted in place of broken edges. Then the homotopy method is applied, each inserted source is swept to find all possible solutions. Any solution other than the designers' given solution is considered a hardware Trojan state. Figure 9.7 presents the application of the proposed technique on the Wilson bias generator. It indicates that there is only one operating point at 0 °C and 50 °C, and there are three operating points between 100 °C ~ 250 °C, and there is only one operating point at 300 °C.

9.2.2 Information Flow Tracking in AMS Circuits

Hardware Trojans in AMS circuit can cross the analog/digital interface or digital/analog interface. A methodology based on proof-carrying hardware intellectual property (PCHIP) has been proposed to enable systematic formal evaluation of information flow policies in AMS circuits [7]. The information flow tracking (IFT)

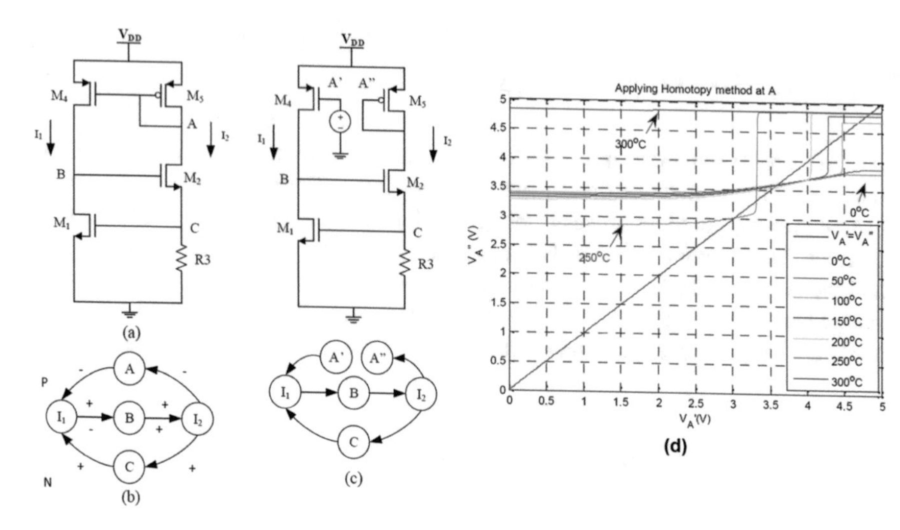

Fig. 9.7 The Wilson bias generator (**a**) circuit, (**b**) directed dependency graph, (**c**) broken Positive feedback loop, and (**d**) Homotopy method over 0–300 °C temperature range [6]

is to extend the transistor level, and a unified framework is developed to enforce information flow policies in digital, analog, and mixed-signal circuits.

Information, e.g. bits of a secret key, in an AMS circuit can be leaked in various ways: (a) information may be carried through both voltage and current; (b) a transistor in an AMS circuit can serve as a common source, common gate, or common drain; thereby, producing gate-to-drain, drain-to-source, or gate-to-source data flows, respectively. Moreover, changing voltage on the source or the drain impacts the drain-to-source current; (c) the voltage on the bulk terminal of a transistor may also be manipulated to leak information to the source or drain terminals; and (d) other components, such as capacitors, resistors, can also involve in information flow. Based on the above observation, some information flow policies for analog circuits have been developed for MOSFETs and bipolar transistors, capacitors, inductors and resistors, and diodes. As a case for bipolar transistors, any value on the base terminal is transferred to the emitter and the collector terminals. And, any sensitive value on the emitter terminal is transferred to the collector and vice versa [7]. Figure 9.8 presents the sample module definitions of analog components. These definitions enable integration of digital and analog circuits in one place to perform information tracking across the domains.

9.2.3 *Side-Channel Fingerprinting*

Hardware Trojan effects on circuit side-channel signals is systematic that is contrary to effects of process and environmental variations. This systematic impact cannot be avoided as the attacker extracts hidden information based on that. For the

```
1 // Modeling analog data flow in capacitors
2 // Resisitors, inductors and diodes are defined similarly
3 module cap (a, b);
4   inout a, b;
5
6   assign a = a & b;
7   assign b = a & b;
8 endmodule
9
10 // Modeling analog data flow in NMOS transistors
11 // PMOS transistors are defined similarly
12 module nch (d, b, g, s);
13   input g;
14   inout d, b, s;
15
16   assign d = d & b & g & s;
17   assign s = d & b & g & s;
18 endmodule
19
20 // Modeling analog data flow in NPN transistors
21 // PNP transistors are defined similarly
22 module npn (b, c, e);
23   input b;
24   inout c, e;
25
26   assign c = b & c & e;
27   assign e = b & c & e;
28 endmodule
```

Fig. 9.8 Sample module definitions to mimic analog data flows in *VeriCoq*-IFT [7]

analog Trojans inserted in the UWB TX [1], the systematic impact results in an extra statistical structural to the transmission power of integrated circuits. It has shown that the principal component analysis (PCA) can effectively reveal the added statistical structural by a hardware Trojan [1].

Assuming the existence of hardware Trojan-free circuit, PCA is performed on 40 hardware Trojan-free circuits, and 40 circuits with hardware Trojan-I and Trojan-II (shown in Fig. 9.1) over randomly selected six different blocks of plaintext that are encrypted through the AES using a randomly chosen 128-bit key. Each of the resulting six blocks of ciphertext was then transmitted by the UWB TX and the total transmission power for each block over the public channel was measured for all circuits. Figure 9.9a clearly indicates that hardware Trojan-free circuits and hardware Trojan-inserted circuit are indistinguishable. However, PCA on the three principal components of the original data could effectively separate and identify the hardware Trojans (Fig. 9.9b).

9.3 Conclusions

This chapter reviewed some of the major studies in hardware Trojans in analog and mix-signal circuits. It has been discussed by different researches that hardware Trojan states inherently exist in AMS circuits, and traditional measures such as startup circuits cannot prevent such hardware Trojans. The design and test of

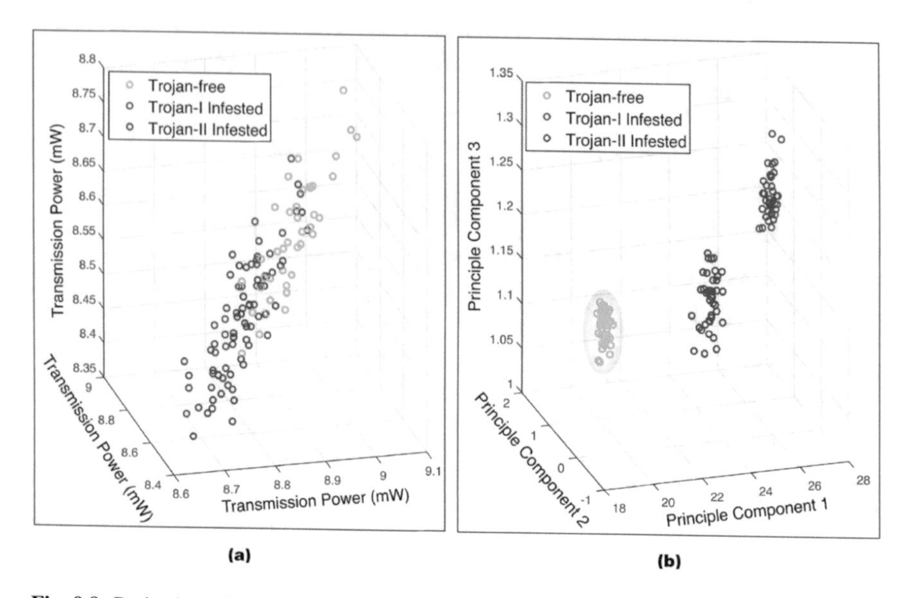

Fig. 9.9 Projection of hardware Trojan-free and hardware Trojan-infested circuits on a 3-D space (**a**) before applying PCA and (**b**) after applying PCA over total transmission power for transmitting one ciphertext block [1]

the AMS circuits demands various expertise in control system, analog circuits, digital signal processing, and devices. Its interdisciplinary nature makes its security analyses very challenging and requires collaboration of various research teams with different backgrounds from academia and industry. While the current state-of-knowledge presents simple types of analog hardware Trojans in very simple AMS circuits, to effectively tackle the analog hardware Trojan issue, there is need to realistic scenarios for the analog hardware Trojan in complex designs. It should be clarified how the trigger and payload of an analog hardware Trojan may look like. It should be carefully studied how analog and digital domains can negatively affect each other and facilitate the realization of analog hardware Trojans. These efforts can lead to development of effective design and testing technique for analog hardware Trojan detection and prevention.

References

1. Y. Liu, Y. Jin, A. Nosratinia, Y. Makris, Silicon demonstration of hardware Trojan design and detection in wireless cryptographic ICs. IEEE Trans. Very Large Scale Integration (VLSI) Syst. **25**(4), 1506–1519 (2017)
2. Q. Wang, R.L. Geiger, D. Chen, Hardware Trojans embedded in the dynamic operation of analog and mixed-signal circuits, in *2015 National Aerospace and Electronics Conference (NAECON)* (2015), pp. 155–158

3. X. Cao, Q. Wang, R.L. Geiger, D.J. Chen, A hardware Trojan embedded in the inverse Widlar reference generator, in *2015 IEEE 58th International Midwest Symposium on Circuits and Systems (MWSCAS)* (2015), pp. 1–4
4. K. Yang, M. Hicks, Q. Dong, T. Austin, D. Sylvester, A2: analog malicious hardware, in *2016 IEEE Symposium on Security and Privacy (SP)* (2016), pp. 18–37
5. G.T. Becker, F. Regazzoni, C. Paar, W.P. Burleson, *Stealthy Dopant-Level Hardware Trojans* (Springer, Berlin, 2013), pp. 197–214
6. Y.T. Wang, Q. Wang, D. Chen, R.L. Geiger, Hardware Trojan state detection for analog circuits and systems, in *NAECON 2014 – IEEE National Aerospace and Electronics Conference* (2014), pp. 364–367
7. M.M. Bidmeshki, A. Antonopoulos, Y. Makris, Information flow tracking in analog/mixed-signal designs through proof-carrying hardware IP, in *Design, Automation Test in Europe Conference Exhibition (DATE)* (2017), pp. 1703–1708

Printed in the United States
By Bookmasters